Advances in Soft Computing 52

Editor-in-Chief: J. Kacprzyk

Advances in Soft Computing

Editor-in-Chief

Prof. Janusz Kacprzyk
Systems Research Institute
Polish Academy of Sciences
ul. Newelska 6
01-447 Warsaw
Poland
E-mail: kacprzyk@ibspan.waw.pl

Further volumes of this series can be found on our homepage: springer.com

Erel Avineri, Mario Köppen, Keshav Dahal,
Yos Sunitiyoso, Rajkumar Roy (Eds.)

Applications of Soft Computing

Updating the State of the Art

 Springer

Editors

Dr. Erel Avineri
Senior Lecturer in Integrated Transport
Centre for Transport & Society
Faculty of the Built Environment
University of the West of England
Frenchay Campus
Coldharbour Lane
Bristol BS16 1QY
UK
E-mail: Erel.Avineri@uwe.ac.uk

Dr. Mario Köppen
Kyushu Institute of Technology
Dept. Artificial Intelligence
680-4 Kawazu
Izuka-Shi, Fukuoka 820-8502
Japan
E-mail: mkoeppen@pluto.ai.kyutech.ac.jp

Dr. Keshav Dahal
MOSAIC Research Group
University of Bradford
Department of Computing
Bradford BD7 1DP
UK
E-mail: K.P.Dahal@Bradford.ac.uk

Yos Sunitiyoso
Centre for Transport & Society
Faculty of the Built Environment
University of the West of England
Frenchay Campus
Coldharbour Lane
Bristol BS16 1QY
UK

Prof. Rajkumar Roy
Professor of Decision Engineering and
Head of Decision Engineering Centre
Manufacturing Department
Cranfield University
Bedford MK43 0AL
UK
Email: r.roy@cranfield.ac.uk

ISBN 978-3-540-88078-3   e-ISBN 978-3-540-88079-0

DOI 10.1007/978-3-540-88079-0

Advances in Soft Computing   ISSN 1615-3871

Library of Congress Control Number: 2008935493

©2009 Springer-Verlag Berlin Heidelberg

Typeset & Cover Design: Scientific Publishing Services Pvt. Ltd., Chennai, India.

Printed in acid-free paper

5 4 3 2 1 0

springer.com

Preface

WSC12 Chair's Welcome Message

Dear Colleague,

On behalf of the Organizing Committee, it is our honor to welcome you to the WSC12 (12th Online World Conference on Soft Computing in Industrial Applications).

Continuing a tradition started over a decade ago by the World Federation of Soft Computing (http://www.softcomputing.org/) to bring together researchers interested in advancing state of the art in the field. Continuous technological improvements since then continue to make this online forum a viable gathering format for a world class conference.

The program committee received a total of 69 submissions from 30 countries, covering all the continents. This reflects the worldwide nature of this event. Each paper was then reviewed by three referees, culminating in the acceptance of 27 papers for publication. Authors of all accepted papers were then notified to prepare and submit their final manuscripts and conference presentations. 26 papers were presented at the conference and are included in this volume.

The organization of the WSC12 conference is completely voluntarily. The review process required an enormous effort from the members of the International Program Committee. We would like to thank all members of the International Program Committee for their contribution to success of this conference. We would like to express our sincere thanks to special session organizers, the plenary presenters, and to the publisher Springer for their hard work and support in organizing the conference. Finally, we would like to thank all the authors for their high quality contributions. It is all of you who make this event possible!

We hope you will enjoy the conference and we are looking forward to meeting you virtually at WSC12. We encourage you to take an active part in the WSC12 paper discussions - your feedback is very important to other authors!

Note on the environment: We would like take this opportunity to raise the awareness to the importance of reducing carbon emissions by participating virtual conferences such as the WSC12. Carbon emissions caused as a result of traveling are one of the major contributors to greenhouse effect and climate change. It is believed by many that telecommuting and virtual conferencing may lead to a significant reduction in carbon emissions even when increased. We hope that by taking part with this and future virtual conferencing we may help in cutting carbon emissions.

February 2008   Erel Avineri
                                                                 Mario Köppen

Welcome Note by the World Federation on Soft Computing (WFSC) Chair Person

On behalf of the World Federation on Soft Computing (WFSC) I would like to welcome you all in the WSC12 cyberspace! The 12th online World Conference on Soft Computing in Industrial Applications provides a unique opportunity for soft computing researchers and practitioners to publish high quality papers and discuss research issues in detail without incurring a huge cost. The conference has established itself as a truly global event on the Internet. The quality of the conference has improved over the years. The WSC12 conference has covered new trends in soft computing to state of the art applications. The conference has also added new features such as virtual exhibition and online presentation.

On the first day of the conference, I wish all the authors, participants and organisers a very informative and useful interaction and virtual experience. Many thanks to the reviewers, sponsors and publishers of WSC12; and congratulations to the organisers! I believe your hard work will make the conference a true success.

October 2007   Rajkumar Roy

WSC12 Best Paper Award

The WSC12 Program announced that the Best Paper Award to honor the author(s) of a paper of exceptional merit dealing with a subject related to the Conference's scope. All papers presented at the WSC12 conference automatically entered the competition. The papers were judged based on the following criteria: The complexity of problem; the originality of the research, the depth of the research; the analysis undertaken; the quality of presentation and the quality of written paper. The feedback received by the reviewers was an important input to the process.

The overall standard of the papers presented at the WSC12 Conference was good, with a small group of 4 or 5 being very strong. We are happy to announce that the paper awarded by the WSC12 Best Paper Award is "A Multi-Agent Classifier System based on the TNC Model" by Anas Quteishat, Chee-Peng Lim, Jeffrey Tweedale and Lakhmi C. Jain.

February 2008

Erel Avineri
Mario Köppen

Organization

Honorary Chair

Lotfi A. Zadeh   University of California Berkeley, USA

General Chair

Erel Avineri   University of the West of England, Bristol, UK

Program Chair

Mario Köppen   Kyushu Institute of Technology, Fukuoka, Japan

International Co-chair

Ashraf Saad   Armstrong Atlantic State University, USA

Publicity Chair

Keshav Dahal   University of Bradford, UK

Web Chair

Yos Sunitiyoso   University of the West of England, Bristol, UK

Chairman of the Advisory Board

Rajkumar Roy   Cranfield University, UK

International Technical Program Committee

Janos Abonyi   University of Veszprem Folyamatmernoki Tanszek, Hungary

Ajith Abraham   Chung-Ang University, Korea

Sudhirkumar V. Barai	Department of Civil Engineering, Kharagpur, India
Valeriu Beiu	United Arab Emirates University, UAE
Christian Blum	Universitat Politecnica de Catalunya, Spain
Zvi Boger	Ben Gurion University of the Negev, Israel
Larry Bull	University of the West of England, Bristol, UK
Oscar Castillo	Instituto Tecnologico de Tijuana, Mexico
Leandro Coelho	Pontifical Catholic University of Parana, Brazil
Carlos A. Coello Coello	CINVESTAV-IPN, Mexico
Oscar Cordon	University of Granada, Spain
Guy De Tré	Ghent University, Belgium
Suash Deb	National Institute of Science & Technology, India
Mauro Dell'Orco	Politecnico di Bari, Italy
Giuseppe Di Fatta	The University of Reading, UK
Takeshi Furuhashi	Nagoya University, Japan
Xiao-Zhi Gao	Helsinki University of Technology, Finland
Antonio Gaspar-Cunha	University of Minho, Portugal
Roderich Gross	Unilever R&D Port Sunlight, UK
Hani Hagras	University of Essex, UK
Ioannis Hatzilygeroudis	University of Patras, Greece
Hisao Ishibuchi	Osaka Prefecture University, Japan
Yaochu Jin	Honda Research Institute Europe, Germany
Uri Kartoun	Ben Gurion University of the Negev, Israel
Frank Klawonn	University of Applied Sciences, Germany
Andreas König	Technische Universitat Kaiserslautern, Germany
Renato Krohling	UFES - Federal University of Espirito Santo, Brazil
Reza Langari	Texas A&M University at Qatar
Luis Magdalena	Universidad Politecnica de Madrid, Spain
Christophe Marsala	Universite P. et M. Currie, France
Patricia Melin	Instituto Tecnologico de Tijuana, Mexico
Sanaz Mostaghim	ETH-Zurich, Switzerland
Mehmet K. Muezzinoglu	University of Louisville, USA
Andreas Nuernberger	Universitat Magdeburg, Germany
Jae C. Oh	Syracuse University, USA
Michele Ottomanelli	Politecnico di Bari, Italy
Vasile Palade	Oxford University, UK
Gerardo Rossel	Universidad Abierta Interamericana, Argentina
Muhammad Sarfraz	King Fahd University of Petroleum and Minerals, Saudi Arabia
Ashutosh Tiwari	Cranfield University, UK
Christopher Turner	Cranfield University, UK
Marley Maria B.R. Vellasco	Pontifical Catholic University of Rio de Janeiro, Brazil
Juan Wachs	Ben Gurion University of the Negev, Israel
Christian Woehler	DaimlerChrysler AG, Germany
Berend Jan van der Zwaag	University of Twente, The Netherlands

WSC12 Technical Sponsors

WSC12 was organized by

WSC12 was sponsored by

WSC12 was technically co-sponsored by

Plenary Presentation Abstracts

Three Applications of Soft Computing and What They Teach Us

Lawrence David Davis

President, VGO Associates
http://www.vgoassociates.com/

Summary. This talk centers on three applications of soft computing–one in pipeline control, one in oil field production, and one in agriculture. Each application used evolutionary computation and other techniques in order to achieve substantial improvements in the operations of the companies involved. In addition to a description of each application, the talk will describe some implications of those successful projects for future practitioners of soft computing as well as companies with problems that soft computing may be used to solve.

Pareto-Based Multi-Objective Machine Learning

Yaochu Jin

Honda Research Institute Europe (HRI-EU), Offenbach, Germany
yaochu.jin@honda-ri.de

Summary. Machine learning is inherently a multi-objective task. Traditionally, however, either only one of the objectives is adopted as the cost function or multiple objectives are aggregated to a scalar cost function. This can mainly attributed to the fact that most conventional learning algorithms can only deal with a scalar cost function. Over the last decade, efforts on solving machine learning problems using the Pareto-based multi-objective optimization methodology have gained increasing impetus, particularly thanks to the great success of multi-objective optimization using evolutionary algorithms and other population-based stochastic search methods. It has been shown that Pareto-based multi-objective learning approaches are more powerful compared to learning algorithms with a scalar cost functions in addressing various topics of machine learning, such as clustering, feature selection, improvement of generalization ability, knowledge extraction, and ensemble generation.

List of Contributors

Janos Abonyi
Department of Process Engineering,
University of Pannonia
Veszprem P.O. Box 158
H-8201 Hungary
abonyij@fmt.uni-pannon.hu

Victor Alchanatis
Institute of Agricultural Engineering
Volcani Center
Bet-Dagan, 50250, Israel
victor@volcani.agri.gov.il

Erel Avineri
Centre for Transport & Society
University of the West of England
Bristol, BS16 1QY, UK
erel.avineri@uwe.ac.uk

Lucia Ballerini
European Centre for Soft Computing
C/ Gonzalo Gutiérrez Quirós s/n
33600 Mieres, Spain
lucia.ballerini@softcomputing.es

S.V. Barai
Indian Institute of Technology
Kharagpur
Kharagpur 721 302, India
skbarai@civil.iitkgp.ernet.in

Gautam Bhattacharya
Deparment of Civil Engineering
Bengal Engineering and Science
University
Howrah-711103, India
bhattacharyag@gmail.com

Gennaro Nicola Bifulco
University of Napoli
via Claudio 21
Napoli, 80125, Italy
gennaro.bifulco@unina.it

Tom Burks
Agricultural and Biological
Engineering
University of Florida
Gainesville,110570, Florida, US
tfburks@ifas.ufl.edu

Deepankar Choudhury
Department of Civil Engineering
Indian Institute of Technology
Bombay, Mumbai - 400076, India
dchoudhury@iitb.ac.in

Vincent Cicirello
The Richard Stockton College of New
Jersey
Computer Science and Information
Systems
Pomona, NJ 08240, USA
cicirelv@stockton.edu

Leandro dos Santos Coelho
Industrial and Systems Engineering
Graduate Program, LAS/PPGEPS,
Pontifical Catholic University of
Paraná, PUCPR,
Imaculada Conceição, 1155, Zip code
80215-901, Curitiba, Paraná, Brazil
leandro.coelho@pucpr.br

Oscar Cordón
European Centre for Soft Computing
C/ Gonzalo Gutiérrez Quirós s/n
33600 Mieres, Spain
orscar.cordon@softcomputing.es

José António Covas
IPC- Institute for Polymers and
Composites / I3N,
University of Minho, Campus de
Azurém
4800-058 Guimarães, Portugal
jcovas@dep.uminho.pt

Keshav Dahal
Modeling Optimization Scheduling
And Intelligent Control (MOSAIC)
Group
Department of Computing, University
of Bradford
Bradford, BD7 1DP, UK
k.p.dahal@bradford.ac.uk

Sergio Damas
European Centre for Soft Computing
C/ Gonzalo Gutiérrez Quirós s/n
33600 Mieres, Spain
sergio.damasi@softcomputing.es

Ivanoe De Falco
Institute of High Performance
Computing and Networking,
National Research Council of Italy
Via P. Castellino 111, 80131 Naples,
Italy
ivanoe.defalco@na.icar.cnr.it

Antonio Della Cioppa
Natural Computation Lab, DIIIE,
University of Salerno,
Via Ponte don Melillo 1, 84084
Fisciano (SA), Italy
adellacioppa@unisa.it

Roberta Di Pace
University of Napoli
via Claudio 21
Napoli, 80125, Italy
roberta.dipace@unina.it

Balazs Feil
Department of Process Engineering,
University of Pannonia
Veszprem P.O. Box 158
H-8201 Hungary
feilb@fmt.uni-pannon.hu

Naizhang Feng
Harbin Institute of Technology at
Weihai
No. 2 Road Wenhuaxi
Weihai, 264209, P.R. China
fengnz@yeah.net

Xiao-Zhi Gao
Helsinki University of Technology
Otakaari 5 A
Espoo 02150, Finland
gao@cc.hut.fi

António Gaspar-Cunha
IPC- Institute for Polymers and
Composites / I3N,
University of Minho, Campus de
Azurém
4800-058 Guimarães, Portugal
agc@dep.uminho.pt

Ruixin Ge
Texas Southern University
3100 Cleburne Avenue, Department
of Transportation Studies, TSU
Houston, TX, 77004, USA
ger@tsu.edu

A.K. Gupta
Indian Institute of Technology
Kharagpur
agupta@civil.iitkgp.ernet.in

Santoso Handri
Nagaoka University of Technology
1603-1 Kamitomiokamachi, Nagaoka
Niigata 940-2124 Japan
bondry@alice.nagaokaut.ac.jp

Alamgir Hossain
Modeling Optimization Scheduling
And Intelligent Control (MOSAIC)
Group
Department of Computing, University
of Bradford
Bradford, BD7 1DP, UK
m.a.hossain1@bradford.ac.uk

Xianlin Huang
Center for Control Theory and
Guidance Technology,
Harbin Institute of Technology,
Harbin, P.R. China
xlinhuang@hit.edu.cn

Kuncup Iswandy
University of Kaiserslautern
Dept. of Electrical and Computer
Engineering
Institute of Integrated Sensor Systems
Erwin-Schroedinger-Str. 12/449
Kaiserslautern, D-67663, Germany
kuncup@eit.uni-kl.de

Lakhmi C. Jain
School of Electrical and Information
Engineering, University of South
Australia
Mawson Lakes Campus, Adelaide
South Australia SA 5095, Australia
lakhmi.jain@unisa.edu.au

Andreas König
University of Kaiserslautern
Dept. of Electrical and Computer
Engineering
Institute of Integrated Sensor Systems
Erwin-Schroedinger-Str. 12/455
Kaiserslautern, D-67663, Germany
koenig@eit.uni-kl.de

Tamas Kenesei
Department of Process Engineering,
University of Pannonia
Veszprem P.O. Box 158
H-8201 Hungary
keneseit@fmt.uni-pannon.hu

Jayachandar Kodali
Indian Institute of Technology
Kharagpur
Kharagpur 721 302, India
jayachandar_k2000@yahoo.co.in

Andrew Koh
Institute for Transport Studies
University of Leeds,
Leeds LS2 9JT, UK
a.koh@its.leeds.ac.uk

Edmond Leung
Aston University
Aston Triangle
Birmingham B4 7ET, UK

Chee-Peng Lim
School of Electrical and Electronic
Engineering, University of Science
Malaysia
Engineering Campus, 14300 Nibong
Tebal
Penang, Malaysia
cplim@eng.usm.my

Xiaoyue Liu
Texas Southern University
3100 Cleburne Avenue, Department
of Transportation Studies, TSU
Houston, TX, 77004, USA
ger@tsu.edu

Liyong Ma
Harbin Institute of Technology at
Weihai
No. 2 Road Wenhuaxi
Weihai, 264209, P.R. China
maliyong@hit.edu.cn

Domenico Maisto
Institute of High Performance
Computing and Networking,
National Research Council of Italy
Via P. Castellino 111, 80131 Naples,
Italy
ivanoe.defalco@na.icar.cnr.it

Viviana Cocco Mariani
Mechanical Engineering Graduate
Program, PPGEM,
Pontifical Catholic University of
Paraná, PUCPR,
Imaculada Conceição, 1155, Zip code
80215-901, Curitiba, Paraná, Brazil

Kazuo Nakamura
Nagaoka University of Technology
1603-1 Kamitomiokamachi, Nagaoka
Niigata 940-2124 Japan
nakamura@kjs.nagaokaut.ac.jp

Lars Nolle
Nottingham Trent University
Clifton Campus
Nottingham NG11 8NS, UK
lars.nolle@ntu.ac.uk

Seppo Ovaska
Utah State University
4120 Old Main Hill
Logan UT 84322-4120, U.S.A.
seppo.ovaska@tkk.fi

Vincenzo Punzo
University of Napoli
via Claudio 21
Napoli, 80125, Italy
vinpunzo@unina.it

Fengxiang Qiao
Texas Southern University
3100 Cleburne Avenue, Department
of Transportation Studies, TSU
Houston, TX, 77004, usa
qiao_fg@tsu.edu

Anas Quteishat
School of Electrical and Electronic
Engineering, University of Science
Malaysia
Engineering Campus, 14300 Nibong
Tebal
Penang, Malaysia
quteishat@gmail.com

Rajkumar Roy
Decision Engineering Centre
Cranfield University
Cranfield, MK43 0AL, UK
r.roy@cranfield.ac.uk

Pijush Samui
Department of Civil Engineering
Indian Institute of Science
Bangalore - 560 012, India
pijush.phd@gmail.com

José Santamaría
Dept. Software Engineering, University of Cádiz
C/ Chile 1
11002 Cádiz, Spain
jose.santamarialopez@uca.es

Umberto Scafuri
Institute of High Performance
Computing and Networking,
National Research Council of Italy
Via P. Castellino 111, 80131 Naples,
Italy
ivanoe.defalco@na.icar.cnr.it

Gerald Schaefer
Aston University
Aston Triangle
Birmingham B4 7ET, UK
g.schaefer@aston.ac.uk

Fulvio Simonelli
University of Napoli
via Claudio 21
Napoli, 80125, Italy
fulsimon@unina.it

Zhuoyue Song
Center for Control Theory and
Guidance Technology,
Harbin Institute of Technology,
Harbin, P.R. China
zhuoyue603@sohu.com

Mark Stelling
Decision Engineering Centre
Cranfield University
Cranfield, MK43 0AL, UK
m.t.stelling@cranfield.ac.uk

Helman Stern
Department of Industrial Engineering
and Management
Ben-Gurion University of the Negev
Beer-Sheva, 84105, Israel
helman@bgu.ac.il

Yude Sun
Harbin Institute of Technology at
Weihai
No. 2 Road Wenhuaxi
Weihai, 264209, P.R. China
sun-yude@163.com

Ernesto Tarantino
Institute of High Performance
Computing and Networking,
National Research Council of Italy
Via P. Castellino 111, 80131 Naples,
Italy
ivanoe.defalco@na.icar.cnr.it

Ashutosh Tiwari
Decision Engineering Centre
Cranfield University
Cranfield, MK43 0AL, UK
a.tiwari@cranfield.ac.uk

Chris Turner
Cranfield University,
Cranfield,
Bedfordshire, UK, MK43 0Al
c.j.turner@cranfield.ac.uk

Jeffrey Tweedale
School of Electrical and Information
Engineering, University of South
Australia
Mawson Lakes Campus, Adelaide
South Australia SA 5095, Australia
jeffrey.tweedale@westnet.com.au

Betzy Varghes
Modeling Optimization Scheduling
And Intelligent Control (MOSAIC)
Group
Department of Computing, University
of Bradford
Bradford, BD7 1DP, UK
betzymol@yahoo.com

Juan Wachs
Department of Industrial Engineering
and Management
Ben-Gurion University of the Negev
Beer-Sheva, 84105, Israel
juan@bgu.ac.il

Xiaolei Wang
Helsinki University of Technology
Otakaari 5 A
Espoo 02150, Finland
xiaolei@cc.hut.fi

Huimin Wang
Center for Control Theory and
Guidance Technology,
Harbin Institute of Technology,
Harbin, P.R. China
happy_wang15@126.com

Lei Yu
Texas Southern University
3100 Cleburne Avenue, Department
of Transportation Studies, TSU
Houston, TX, 77004, USA
yu_lx@tsu.edu

Valerio de Martinis
University of Napoli
via Claudio 21
Napoli, 80125, Italy
vdemartinis@unina.it

Contents

Part III: Signal Processing and Pattern Recognition

Part IV: Civil Engineering

Part V: Computer Graphics, Imaging and Vision

## Part I

Design of Soft Computing Techniques

Linguistic Information in Dynamical Fuzzy Systems — An Overview

Xiao-Zhi Gao[1], Seppo Ovaska[2], and Xiaolei Wang[1]

[1] Department of Electrical Engineering, Helsinki University of Technology,
Otakaari 5 A, FI-02150 Espoo, Finland
[2] Department of Electrical and Computer Engineering, Utah State University,
4120 Old Main Hill, Logan, UT 84322-4120 U.S.A.

Summary. This paper gives an overview on the study of using linguistic information in the dynamical fuzzy systems. We introduce the Linguistic Information Feed-Back-based Dynamical Fuzzy System (LIFBDFS) and Linguistic Information Feed-Forward-based Dynamical Fuzzy System (LIFFDFS), in which the past fuzzy inference output represented by a membership function is fed back and forward, respectively. Their principles, structures, and learning algorithms are also presented. Both the LIFBDFS and LIFFDFS can overcome the common static mapping drawback of conventional fuzzy systems. They have the distinguished advantage of inherent dynamics, and are, therefore, well suited for handling temporal problems, such as process modeling and control. Our survey paper provides a general guideline for interested readers to choose, design, and apply these linguistic information feed-back- and feed-forward-based dynamical fuzzy systems in engineering.

1 Introduction

Due to its great flexibility in coping with ill-defined plants, fuzzy logic theory has found numerous successful applications in industrial engineering, e.g., pattern recognition [1], automatic control [2], and fault diagnosis [3]. Generally, a fuzzy logic-based system with embedded linguistic knowledge maps an input (feature) vector to a scalar (conclusion) output [2]. Most fuzzy systems deployed in practice are static. In other words, they lack the necessary internal dynamics, and can, thus, only implement nonlinear but non-dynamical input-output mappings. This disadvantage greatly hinders their wider employment in such areas as dynamical system modeling, prediction, signal filtering, and control. Inspired by the idea of utilizing linguistic information, we have proposed two kinds of dynamical fuzzy systems, Linguistic Information Feed-Back-based Dynamical Fuzzy System (LIFBDFS) [4] and Linguistic Information Feed-Forward-based Dynamical Fuzzy System (LIFFDFS) [5]. Instead of crisp system output, conclusion fuzzy membership function is fed back and forward locally with adjustable parameters in the LIFBDFS and LIFFDFS, respectively. Taking full use of the available linguistic information, these two novel dynamical fuzzy systems can overcome certain drawbacks of the regular fuzzy systems, and are considered as attractive choices for dynamical system identification and control. This

E. Avineri et al. (Eds.): Applications of Soft Computing, ASC 52, pp. 3–12.
springerlink.com   © Springer-Verlag Berlin Heidelberg 2009

paper summarizes the recent progresses in employing the linguistic information to build the dynamical fuzzy systems. The principles, structures, as well as learning algorithms of both the LIFBDFS and LIFFDFS are briefly presented and discussed. The aim of our overview is to give the readers some helpful insights in the state-of-the-art linguistic information feed-back- and feed-forward-based dynamical fuzzy systems.

Our paper is organized as follows: the LIFBDFS is first introduced in Section 2. The basic and generalized structures of the LIFBDFS are also discussed here. In the next section, a simplified LIFBDFS model, S-LIFBDFS, is further presented. The underlying principle of the LIFFDFS is given in Section 4. Finally, in Section 5, we conclude this paper with a few conclusions and remarks.

2   Linguistic Information Feed-Back-Based Dynamical Fuzzy System (LIFBDFS)

2.1   Basic Structure of LIFBDFS

To implant internal dynamics into static fuzzy systems, a common solution is globally feeding the defuzzified (crisp) output to the input through a unit delay [6]. However, in such recurrent fuzzy systems, the linguistic information of the conclusions drawn from fuzzy inference is not fully utilized. These useful knowledge that can reflect the dynamical behaviors of fuzzy systems is unavoidably lost during the defuzzification phase. The conceptual structure of our Linguistic Information Feed-Back-based Dynamical Fuzzy System (LIFBDFS) is shown in Fig. 1.

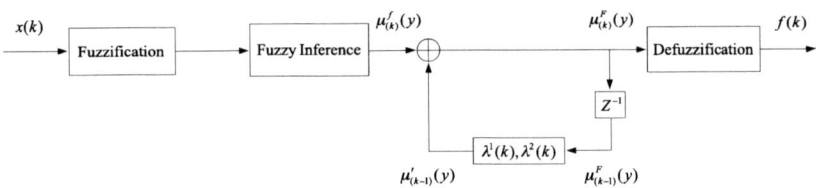

Fig. 1. Structure of Linguistic Information Feedback-based Dynamical Fuzzy System (LIFBDFS)

It is clearly visible at iteration k, the one step delayed conclusion in term of a fuzzy membership function, $\mu^F_{(k-1)}(y)$, is fed back to the output directly drawn from inference rules, $\mu^f_{(k)}(y)$. From the structure point of view, the LIFBDFS with this linguistic information feed-back loop inside can be considered as a locally recurrent neural network. We emphasize that only the Mamdani-type fuzzy systems are considered here. Two adaptive coefficients, i.e., $\lambda^1(k)$ and $\lambda^2(k)$, are employed to adjust the height, width, and center location of the feed-back membership function $\mu^F_{(k-1)}(y)$. If the *max* T-conorm is used for aggregation, $\mu^F_{(k)}(y)$ is obtained:

$$\mu_{(k)}^{F}(y) = \max \left[\mu_{(k)}^{f}(y), \mu_{(k-1)}^{f}(y) \right] \qquad (1)$$

Generally, the defuzzification operation is then applied to acquire a crisp value from $\mu_{(k)}^{F}(y)$. Suppose $f(k)$ is the defuzzified output of our LIFBDFS, and we get:

$$f(k) = \text{DEFUZZ}\left[\mu_{(k)}^{F}(y) \right], \qquad (2)$$

where $\text{DEFUZZ}\left[\,\right]$ is a defuzzification operator. An illustrative diagram of the fuzzy information feed-back in the LIFBDFS is shown in Fig. 2.

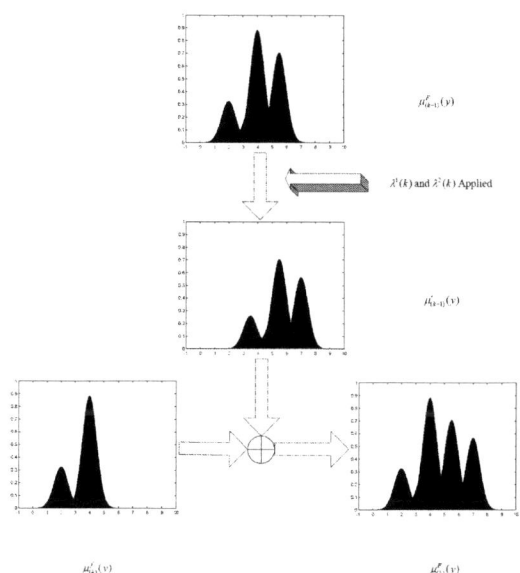

Fig. 2. Diagram of linguistic information feed-back in LIFBDFS

In a typical function approximation or system identification case, if $f_d(k)$ is denoted as the desired output of the LIFBDFS at k, we can define the following approximation error:

$$E(k) = \frac{1}{2} \left[f_d(k) - f(k) \right]^2. \qquad (3)$$

Therefore, the goal of the LIFBDFS learning algorithm is to train those feed-back parameters so that $E(k)$ can be minimized. Actually, λ^1 and λ^2 are intuitively considered as gains in the feedback loop of the LIFBDFS, which makes it possible to utilize *a priori* knowledge to choose their appropriate initial values.

2.2 Structure Extension of LIFBDFS

In Fig. 1, the basic LIFBDFS only feeds back the single-step delay final conclusion output, $\mu^F_{(k-1)}(y)$, inferred from all the fuzzy rules. However, the idea of
linguistic information feedback can be directly generalized to *individual* rules,
as illustrated in Fig. 3, where N is the number of fuzzy rules. Note, each rule
here has a local and independent linguistic feedback. Compared with the two
feedback coefficients case in Fig. 1, there are totally $2 \times N$ trainable parame-
ters now, i.e., $\{\lambda^1_1, \lambda^2_1, \lambda^1_2, \lambda^2_2, \cdots, \lambda^1_N, \lambda^2_N\}$, for these rules. From the adaptation
point of view, more degrees of freedom have been introduced in this extended
LIFBDFS structure.

Besides the first-order linguistic information feedback structures in Figs. 1
and 3, we can also construct higher-order feedback structures. As shown in
Fig. 4, for Rule l, the multi-step delayed membership functions $\mu^f_{(k-1)}(y)$,
$\mu^f_{(k-2)}(y), \cdots, \mu^f_{(k-M)}(y)$ (M is the feedback order) are fed back in separate
loops. Hence, the number of feedback parameters for only a single rule, $\{\lambda^1_1(k),$
$\lambda^2_1(k), \lambda^1_2(k), \lambda^2_2(k), \cdots, \lambda^1_M(k), \lambda^2_M(k)\}$, is $2 \times M$. In case of N rules again, there
are $2 \times N \times M$ coefficients altogether. History information in the input signal
is saved by these loops. This high-order linguistic information feedback configuration can significantly enhance the dynamics of our LIFBDFS, which results
in a proper candidate well suited for coping with temporal problems. Detailed
discussions of more structure extensions of our LIFBDFS can be found in [4] [5].
Practical LIFBDFS should combine the structures demonstrated in both Figs. 3
and 4.

2.3 Learning Algorithm of LIFBDFS

The proposed LIFBDFS has several distinguished features. Firstly, pure fuzzy
inference output is fed back without any information transformation and loss.
Secondly, the local feedback loops act as internal memory units, and can, thus,

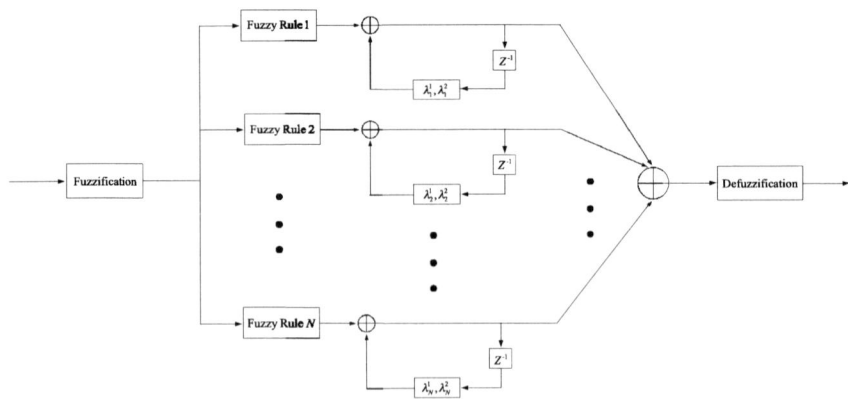

Fig. 3. Linguistic information feedback for individual fuzzy rules in LIFBDFS

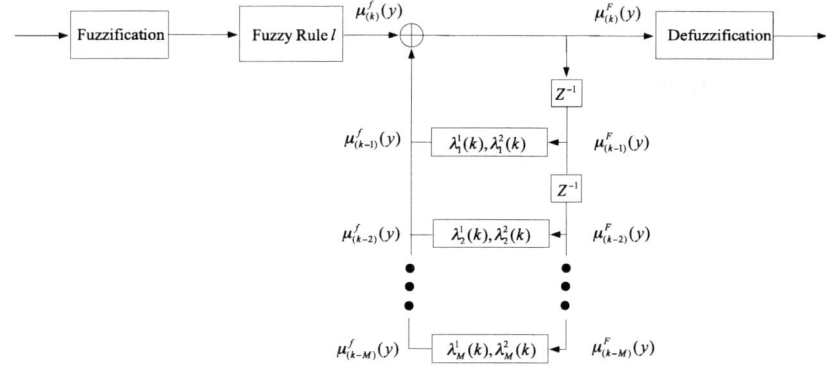

Fig. 4. M-order linguistic information feedback in LIFBDFS

store temporal sequences of input signals, which is pivotal in modeling dynamical systems. Finally, training the feedback coefficients leads to an equivalent update of those fuzzy output membership functions. As a matter of fact, this parameter refinement approach adds the advantage of self-adaptation to our LIFBDFS.

Unfortunately, deriving an explicit and analytical learning algorithm for the LIFBDFS is rather difficult, because of not only its unique recurrent structure but also strong nonlinearity of the linguistic information feedback. Some general-purpose algorithmic optimization schemes, e.g., Genetic Algorithms (GA) [7], could be deployed in the first place. Nevertheless, it is difficult to apply such GA-based training methods for *on-line* applications, since they are not computationally incremental, and often converge very slowly. In [8], the α-level sets representation of fuzzy membership functions and Backpropagation Through Time (BTT) technique are employed to derive a closed-form learning algorithm for the feedback parameters of the LIFBDFS. With this training algorithm, our LIFBDFS can be used to solve real-time modeling and prediction problems.

For the purpose of applying real number arithmetic operations to train the feedback parameters in the LIFBDFS, we here use the α-level sets of fuzzy numbers to represent the linguistic information. It is well known that fuzzy numbers are regarded as crisp sets of real numbers, which are contained within different intervals with certain degrees. Thus, they can be defined by the families of their α-level sets according to the resolution identity theorem [9]. More precisely, a fuzzy number, $\mu(x)$, is decomposed into its α-level sets:

$$\mu(x) = \bigcup_{\alpha} \alpha \left[a^{(\alpha)}, b^{(\alpha)} \right], \tag{4}$$

where $\alpha \in [0,1]$, and $a \in x$ and $b \in x$ with $\mu[a^{(\alpha)}] = \mu[b^{(\alpha)}] = \alpha$. α refers to the corresponding membership grade of the interval $[a^{(\alpha)}, b^{(\alpha)}]$. Especially, the extension principle also applies to these operations. The benefit of using the α-level sets to represent fuzzy numbers is that the operations on fuzzy numbers are

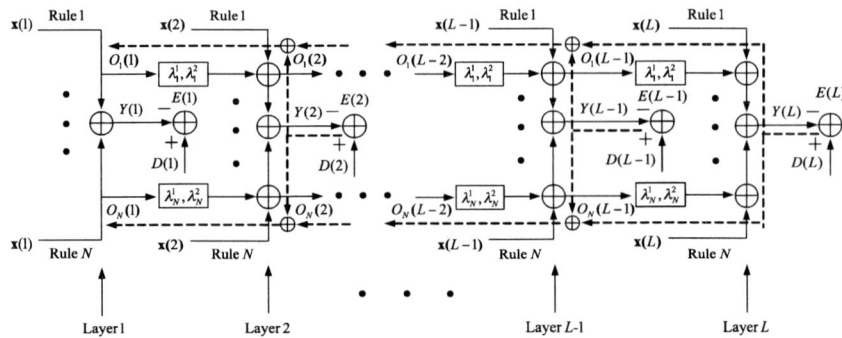

Fig. 5. LIFBDFS unfolded of time in L duplications

accomplished by numerical calculations based on the extension principle, since the α-level sets are actually crisp sets. Therefore, linguistic information can be manipulated conveniently under an explicit real number operation framework. That is, all the existing operations in the LIFBDFS including fuzzification, inference, aggregation, and defuzzification are indeed performed by the interval arithmetic operations for α-level sets [9].

As we know, the training problem of recurrent neural networks is usually solved by using Werbos' Backpropagation Through Time (BTT) approach [10]. In other words, the recurrent neural networks are first unfolded of time to transform them into the feedforward ones, and their output errors are then back-propagated through every layer as usual to update the connection weights. Inspired by this idea, in order to derive a learning algorithm for the feedback parameters of the LIFBDFS, we can unfold an n-input-one-output LIFBDFS into L equivalent copies, as illustrated in Fig. 5, where L is the number of time intervals in duplication, and $x(k) = [x_1(k), x_2(k), \cdots, x_n(k)]$ and $Y(k)$ are the input and output of the LIFBDFS in Layer k, respectively. Thus, based on the above transformed feedforward structure as well as α-level sets representation of the linguistic information, the regular gradient descent principle [11] can be applied to acquire the learning algorithm for our LIFBDFS. More relevant details are given in [8]. Unfortunately, the above LIFBDFS learning algorithm has a considerably high computational complexity. In the next section, we introduce a Simplified LIFBDFS (S-LIFBDFS) model [12]. Compared with the original LIFBDFS, this S-LIFBDFS only utilizes partial linguistic information for feedback with moderately weaker nonlinearity.

3   Simplified Linguistic Information Feed-Back-Based Dynamical Fuzzy System (S-LIFBDFS)

As we can observe from Fig. 2, the effect of feedback parameters $\lambda^1(k)$ and $\lambda^2(k)$ is implicitly embedded in the transformation from $\mu_{(k-1)}^F(y)$ to $\mu'_{(k-1)}(y)$, and the major nonlinearity of linguistic information feedback lies in the aggregation of

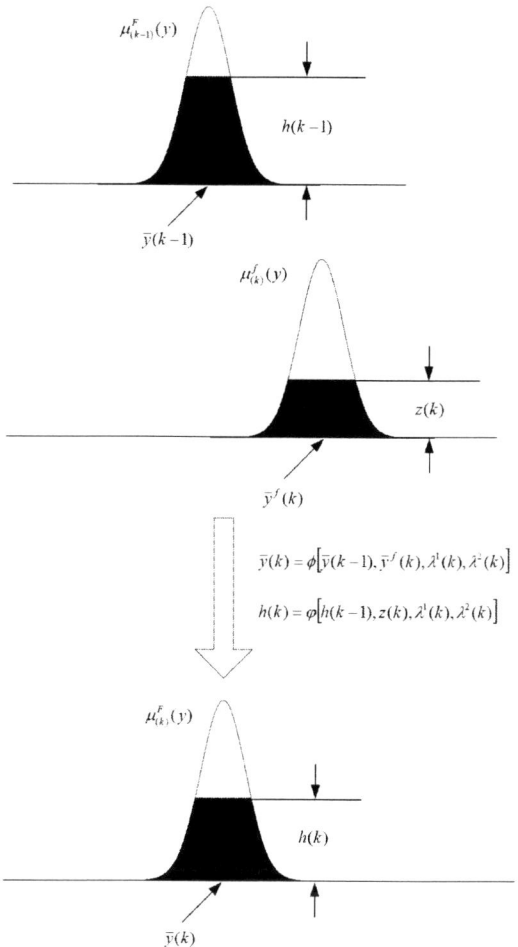

Fig. 6. Simplified linguistic information feedback

feed-back membership function $\mu'_{(k-1)}(y)$ and current inference output $\mu^f_{(k)}(y)$. With all the fuzzy information contained, $\mu'_{(k-1)}(y)$ significantly changes $\mu^f_{(k)}(y)$ so that $\mu^F_{(k)}(y)$ is finally obtained. These facts actually make it rather difficult to derive an explicit learning algorithm for the feedback parameters. Therefore, we propose a Simplified LIFBDFS model: S-LIFBDFS, in which only the center location and height of $\mu^f_{(k)}(y)$ are adjusted by the characteristics information extracted from $\mu^F_{(k-1)}(y)$, while the shape of $\mu^f_{(k)}(y)$ is always kept unchanged [12]. Figure 6 illustrates the diagram of our simplified linguistic information feedback. $\mu^F_{(k-1)}(y)$, $\mu^f_{(k)}(y)$, and $\mu^F_{(k)}(y)$ are denoted by the corresponding shaded areas of the membership functions. In any fuzzy rule, let $\bar{y}(k-1)$ and $h(k-1)$ represent the center and height of the feedback membership function $\mu^F_{(k-1)}(y)$, respectively, at iteration k. For the current concluded membership function $\mu^f_{(k)}(y)$, $\bar{y}^f(k)$ and $z(k)$ are its center and height, respectively. Instead of taking advantage of the whole $\mu^F_{(k-1)}(y)$ to change $\mu^f_{(k)}(y)$ in the LIFBDFS, we consider $\bar{y}(k-1)$ and $h(k-1)$ to be the characteristics representation of $\mu^F_{(k-1)}(y)$, and employ only these two parameters as the 'feed-back information'. In other words, the center and height of $\mu^F_{(k)}(y)$, $\bar{y}(k)$ and $h(k)$, are completely determined by $\bar{y}(k-1)$ & $\bar{y}^f(k)$ and $h(k-1)$ & $z(k)$ as well as feedback parameters $\lambda^1(k)$ & $\lambda^2(k)$. Obviously, the shape of $\mu^F_{(k)}(y)$ is not affected by $\mu^F_{(k-1)}(y)$, which is the same as that of $\mu^f_{(k)}(y)$. Therefore, $\bar{y}(k)$ and $h(k)$ can be calculated:

$$\bar{y}(k) = \phi\left[\bar{y}(k-1), \bar{y}^f(k), \lambda^1(k), \lambda^2(k)\right], \qquad (5)$$

$$h(k) = \varphi\left[h(k-1), z(k), \lambda^1(k), \lambda^2(k)\right], \qquad (6)$$

where $\phi[]$ and $\varphi[]$ are two general functions. Based on (5) & (6) and (1), it is apparent that without the full membership function $\mu^F_{(k-1)}(y)$ involved, the original principle of linguistic information feedback is still partially utilized in the S-LIFBDFS. However, with such a simplified linguistic information feedback structure, the learning algorithm of the S-LIFBDFS can be derived using the gradient descent method. Numerical simulations demonstrate the effectiveness of our S-LIFBDFS together with the training algorithm in time series prediction [13].

4   Linguistic Information Feed-Forward-Based Dynamical Fuzzy System (LIFFDFS)

As discussed in Section 2, it is difficult, if not impossible, to derive an analytical learning algorithm for the LIFBDFS, because of the strong nonlinearity in the linguistic information feedback. Additionally, the LIFBDFS may suffer from running instability with inappropriately trained feedback parameters. Inspired by the idea of linguistic information feedback, we proposed a new Linguistic Information Feed-Forward-based Dynamical Fuzzy System (LIFFDFS), in which the past fuzzy inference output is fed forward locally with trainable parameters [5]. Like the LIFBDFS, this LIFFDFS retains the advantage of inherent dynamics. However, in contrast to the LIFBDFS, the corresponding parameter training algorithm can be obtained in a straightforward way.

The conceptual structure of our LIFFDFS (first-order) is shown in Fig. 7. The one step-delayed inference output in terms of a fuzzy membership function, $\mu^f_{(k)}(y)$, is locally fed forward to the final LIFFDFS output (before defuzzification), $\mu^F_{(k+1)}(y)$. Similarly with the LIFBDFS, this structure can be generalized to the multiple rule as well as higher-order linguistic information feed-forward cases. α and β are the 'gains' in the feed-forward loop of the LIFFDFS. Training the feed-forward coefficients is an implicit adjustment of those fuzzy membership

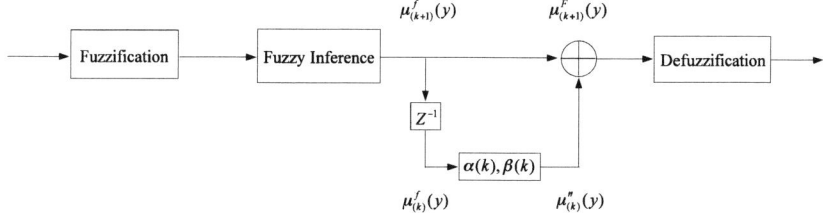

Fig. 7. Structure of LIFFDFS (first-order)

functions of the input variables. On the other hand, different from the LIFBDFS, the LIFFDFS does not have any potential instability problem, due to its unique feed-forward structure. An efficient learning algorithm for our LIFFDFS based on Wang's work [14] and gradient descent principle has been derived [5]. It is also evaluated using the well-known Box-Jenkins gas furnace data [15]. Simulation results show that this new dynamical fuzzy system is a suitable candidate for coping with dynamical problems.

5   Conclusions and Further Work

In this paper, we give a concise overview of utilizing the linguistic information in incorporating dynamics into the static fuzzy systems. A few new dynamical fuzzy systems, i.e., LIFBDFS, S-LIFBDFS, and LIFFDFS, are briefly introduced. Their underlying principles, structures, and learning algorithms are also presented. Although only the Mamdani-type fuzzy models are discussed here, dynamical Sugeno-type fuzzy systems can be similarly constructed with the embedded linguistic information feed-back and feedforward. There are still several challenging problems in the study of both the LIFBDFS and LIFFDFS, such as analytic stability analysis. Moreover, applications of these dynamical fuzzy systems together with their learning algorithms in dealing with real-world engineering problems should be investigated.

References

1. Pedrycz, W.: Fuzzy sets in pattern recognition: Methodology and methods. Pattern Recognition 23(1), 121–146 (1990)
2. Lee, C.C.: Fuzzy logic in control systems: Fuzzy logic controller, Parts I and II. IEEE Transactions on Systems, Man, and Cybernetics 20(2), 404–435 (1990)
3. Kwong, W.A., Passino, K.M., Laukonen, E.G., Yurkovich, S.: Expert supervision of fuzzy learning systems for fault tolerant aircraft control. Proc. IEEE 83(3), 466–483 (1995)
4. Gao, X.Z., Ovaska, S.J.: Dynamical fuzzy systems with linguistic information feedback. In: Melo-Pinto, P., Teodorescu, H.-N., Fukuda, T. (eds.) Systematic Organization of Information in Fuzzy Systems. IOS Press, Amsterdam (2003)

5. Gao, X.Z., Ovaska, S.J.: Linguistic information feed-forward-based dynamical fuzzy systems. IEEE Transactions on Systems, Man, and Cybernetics, Part C 36(4), 453–463 (2006)
6. Zhang, J., Morris, A.J.: Recurrent neuro-fuzzy networks for nonlinear process modeling. IEEE Transactions on Neural Networks 10(2), 313–326 (1999)
7. Tang, K.S., Man, K.F., Kwong, S., He, Q.: Genetic algorithms and their applications. IEEE Signal Processing 13(6), 22–37 (1996)
8. Gao, X.Z., Ovaska, S.J., Wang, X.: Learning algorithm for Linguistic Information Feedback-based Dynamical Fuzzy Systems (LIFDFS). In: Proc. IEEE International Conference on Systems, Man, and Cybernetics, The Hague, The Netherlands, October 2004, pp. 2278–2285 (2004)
9. Lin, C.-T., Lu, Y.-C.: A neural fuzzy system with linguistic teaching signals. IEEE Transactions on Fuzzy Systems 3(2), 169–188 (1995)
10. Werbos, P.J.: Backpropagation through time: What it does and how to do it. Proc. IEEE 78(10), 1550–1560 (1990)
11. Haykin, S.: Neural Networks: A Comprehensive Foundation, 2nd edn. Prentice-Hall, Upper Saddle River (1999)
12. Gao, X.Z., Ovaska, S.J., Wang, X.: A simplified linguistic information feedback-based dynamical fuzzy system with learning algorithm — Part I: Theory. In: Proc. IEEE Mid-Summer Workshop on Soft Computing in Industrial Applications, Espoo, Finland, June 2005, pp. 44–50 (2005)
13. Gao, X.Z., Ovaska, S.J., Wang, X.: A simplified linguistic information feedback-based dynamical fuzzy system with learning algorithm — Part II: Evaluation. In: Proc. IEEE Mid-Summer Workshop on Soft Computing in Industrial Applications, Espoo, Finland, June 2005, pp. 51–56 (2005)
14. Wang, L.-X.: A Course in Fuzzy Systems and Control. Prentice-Hall, Upper Saddle River (1997)
15. Box, G.E.P., Jenkins, G.M.: Time Series Analysis: Forecasting and Control, 2nd edn. Holden-Day, San Francisco (1976)

PSO-Optimized Negative Selection Algorithm for Anomaly Detection

Huimin Wang[1], X.Z. Gao[2], Xianlin Huang[1], and Zhuoyue Song[1]

[1] Center for control theory and guidance technology, Harbin Institute of Technology,
Harbin, P.R. China
`happy_wang15@126.com, xlinhuang@hit.edu.cn, zhuoyue603@sohu.com`
[2] Institute of Intelligent Power Electronics, Helsinki University of Technology, Espoo, Finland
`gao@cc.hut.fi`

Abstract. In this paper, the Particle Swarm Optimization (PSO) is used to optimize the randomly generated detectors in the Negative Selection Algorithm (NSA). In our method, with a certain number of detectors, the coverage of the non-self space is maximized, while the coverage of the self samples is minimized. Simulations are performed using both synthetic and real-world data sets. Experimental results show that the proposed algorithm has remarkable advantages over the original NSA.

Keywords: Anomaly detection, artificial immune system, negative selection algorithm, particle swarm optimization, self-nonself discrimination.

1 Introduction

The Artificial Immune System (AIS) has recently drawn significant research attention from different communities. Forrest and her group at the University of New Mexico have successfully applied the immunological principles in computer security [1]. They explore the potential applications of the AIS in anomaly detection of time series data [2]. The Negative Selection Algorithm (NSA) is inspired by the negative selection mechanism of the natural immune system, which discriminates between the self and non-self. The NSA can improve the accuracy of the conventional detection methods. The proposed anomaly detection technique is considered to be robust, extendible, collateral, and adaptive to detect unknown computer virus attacks.

Currently, the most important problem needed to be solved urgently in the NSA for anomaly detection is the trade-off between the number of detectors and the coverage of non-self space by detectors, that is, how to realize more coverage of non-self space using fewer detectors. Fabio González [3] used simulated annealing to find a good distribution of the detectors that maximizes the coverage of the non-self space.

This paper proposes a novel NSA for anomaly detection. In order to maximize the coverage of the non-self space as well as minimize the coverage of the self samples, the PSO is used to optimize the randomly generated detectors in the original NSA. The remainder of our paper is organized as follows: Section 2 discusses the preliminaries of the NSA with applications in anomaly detection. Section 3 introduces the principles of the PSO and PSO-based NSA detectors optimization. Section 4 demonstrates the computer simulations results. In Section 5, some conclusions and remarks are given.

E. Avineri et al. (Eds.): Applications of Soft Computing, ASC 52, pp. 13–21.
springerlink.com   © Springer-Verlag Berlin Heidelberg 2009

2 Negative Selection Algorithm (NSA) in Anomaly Detection

2.1 Negative Selection Algorithm (NSA)

Forrest *et al.* develop the NSA based on the principles of self-nonself discrimination in the biological immune system [5]. The NSA can be summarized as follows:

1) Define the self as a collection of strings of length *l* over a finite alphabet, a collection that needs to be monitored. For example, *S* can be a normal pattern of computer network activity, which is segmented into equally-sized substrings.

2) Generate a set of detectors *R*, each of which fails to match any string in *R*.

3) Monitor *S* for changes by continually matching the detectors in *R* against *S*. If any detector matches, a change is known to have occurred, because the detectors have been previously designed not to match any of the original strings in *S*.

In the original NSA, elements in the shape space and detectors (matching rules) are represented by binary strings. However, binary representation [4] fails to capture the structure of even simple solution spaces. In addition, the binary matching rules also cause certain scalability problems, because the number of detectors needed grows exponentially with the size of the self. Gonalez *et al.* propose a new NSA using the real-valued representation (RNSA) [6]. In the RNSA, data is expressed by the structure of the higher-level representation, which can significantly speed up the detector generation process. Particular matching rules are associated to the detectors, typically, based on a Euclidean distance. The RNSA has been employed for the aircraft fault detection with satisfactory performances by Dasgupta *et al* [7]. So, RNSA is used in this paper to generate the initial detectors. Section 2.2 describes it in detail.

2.2 NSA for Anomaly Detection

Some definitions used in this paper are given as follows: all the system states are represented by a set *U*, which is composed of *n*-dimensional real-valued vectors. A self set *S* can represent the subset of the states that are considered as normal for the system, and *S* is the subset of *U*. Its complementary set is called non-self set *N*, and is defined as *N*=*U*-*S*. The space covered by *N* is the non-self space. The detector set *D* consists of *n*-dimensional real-valued vectors, each of which fails to match any element in *S*. As a matter of fact, *D* is the subset of *N*, i.e., $D \subseteq N$.

2.2.1 Data Processing and Encoding

The time series data under investigation by the Negative Selection Algorithm is first normalized to the interval [0, 1]. Next, self samples are generated from the time series $X = \{x_1,...x_i,...,x_l\}$ in overlapped sliding window $Y = \{(x_1,...,x_n), (x_2,...,x_{n+1}),..., (x_{l-n+1},...,x_l)\}$, where *n* is the window size. $d(x,y)$ is defined as the Euclidean distance in the space of $[0.0,1.0]^n$, $d(x, y) = \left(\sum_{i=1}^{n} |x_i - y_i|^2 \right)^{1/2}$.

2.2.2 Generation of Detectors

Each self element and NSA detector can be represented by a center and a radius, that is, self $s=(c_s, r_s)$, and detector $d=(c_d, r_d)$, where c_s and c_d are *n*-dimensional points that correspond to the center of hyper-spheres with r_s and r_d as their radii, respectively. The dimension n is equivalent to the window size. Note all the self elements have the same radius $r_s=Rs$. r_d is a variable parameter, which is used in the variable-sized detector (V-detector) (RNSA) [8,9]. In comparison with the RNSA with constant-sized detectors, the V-detector approach is more effective using fewer detectors with variable sizes. It provides a more concise representation of the detection rules learned from training data. The 'holes' in the non-self space can be effectively covered, because smaller detectors are more acceptable, when larger detectors are used to cover the large non-self region.Generation of detectors by the NSA can be described as follows: suppose vector x_r is a random sample from $[0.0,1.0]^n$, if $\min(d(x_r, c_s))<Rs\in R^+$, then $x_r \in S$, and x_r is deleted. Otherwise, estimate whether x_r falls into the detectors that have been already generated. If $d(x_r,c_d)<r_d$, then x_r is deleted. Otherwise, x_r is added to the detector set D as a new detector, the radius of which is $r= \min(d(x_r, c_s))- Rs$.

2.2.3 Anomaly Detection

The principle of anomaly detection in time series using the NSA is explained. Given $\forall x^j$, if $\exists d_i=(c^i_d, r_i)$ satisfies $d(x^j, c^i_d,)< r_i$, then $x^j \in N$, and the time series segment is abnormal. Otherwise, $x^j\in S$, and the time series segment is normal. In order to examine the effectiveness of anomaly detection techniques, detection rate DR and false alarm rate FAR are defined: $DR=TP/(TP+FN)$ and $FAR=FP/(TN+FP)$, where TP stands for the anomalous elements identified as anomalous, FP stands for the normal elements identified as anomalous, FN stands for the anomalous elements identified as normal, and TN stands for the normal elements identified as normal.

3   Optimization of NSA Detectors Using PSO

In the NSA for anomaly detection, coverage of the non-self space by the detectors is a key issue, i.e., how to maximize the coverage of the non-self space using the smallest number of detectors. In this paper, the PSO is employed to optimize the randomly generated detectors by the original NSA. The underlying idea is to expand the coverage of the non-self space under the constraint that the coverage of the self samples should be as small as possible, and the number of the detectors is fixed.

3.1 Particle Swarm Optimization (PSO)

The PSO is a population-based stochastic optimization technique developed by Eberhart and Kennedy in 1995 [10], inspired by the social behaviors of bird flocking and fish schooling. Unlike the Genetic Algorithms (GA), the PSO has no evolution operators, such as crossover and mutation. In the PSO, the potential solutions, namely particles, fly through the solution space by following the current optimum particles. During the past decade, the PSO has been successfully applied in numerous application areas. It is demonstrated that the PSO can achieve better optimization results in a faster manner compared with other classical optimization methods.

3.2 Objective Function

As aforementioned, the goal of detector optimization in the NSA is to maximize the coverage of the non-self space. Therefore, the objective function is the volume of detectors in (1), and the constraint is that detectors cannot fall into the self space:

$$C(D) = Volume(D) \quad s.t. D \not\subset self, \tag{1}$$

where $C(D)$ is the objective function, and $Volume(D)$ is the detector space. In general, the volume of a hyper-sphere is given as $Volume_n(R) = \dfrac{(\pi/2)^k (2R)^n}{n!!}$. The calculation of the volume inside the boundaries of the unit hypercube is a challenging problem in the case of edge hyper-spheres. Additionally, the estimation of the overlapped volume of detectors is rather difficult, especially, when different hyper-shape detectors are considered. To maximize the coverage of the non-self space is actually to reduce the overlapping of detectors. These constraints can be embedded into the objective function, which is described in (2):

$$C(D) = Self \, covering(D) + Overlapping(D) + W * \sum_{s_j} \sum_{d_i} \left\| c_d^i - c_S^j \right\|^2, \tag{2}$$

where $Overlapping(D)$ is the overlapping among the NSA detectors, and $selfcovering(D)$ is the coverage of the self space by these detectors. $W * \sum_{s_j} \sum_{d_i} \left\| c_d^i - c_S^j \right\|^2$ is the weighted distance between the detectors and self set, which controls the distribution of detectors. W is a problem-dependant weight.

Here, the overlapping between the ith detector d_i and j^{th} detector d_j can be approximatively defined as:

$$Overlapping(d_i, d_j) = \begin{cases} 0 & if \ \left\| c_d^i - c_d^j \right\| \geq r_d^i + r_d^j \\ (\exp(\dfrac{r_d^i + r_d^j - \left\| c_d^i - c_d^j \right\|}{r_d^i + r_d^j}) - 1)^n & if \ \left\| c_d^i - c_d^j \right\| < r_d^i + r_d^j \end{cases} \tag{3}$$

$$Overlapping(D) = \sum_{i \neq j} Overlapping(d_i, d_j). \tag{4}$$

Similarly, the overlapping between detector d_i and self s_j is defined as:

$$Self \, covering(d_i, s_j) = \begin{cases} 0 & if \ \left\| c_d^i - c_S^j \right\| \geq r_d^i + r_S^j \\ (\exp(\dfrac{r_d^i + r_S^j - \left\| c_d^i - c_S^j \right\|}{r_d^i + r_S^j}) - 1)^n & if \ \left\| c_d^i - c_S^j \right\| < r_d^i + r_S^j \end{cases} \tag{5}$$

$$Self \, cov \, ering \, (D) = \sum_{s_j} \sum_{d_i} Self \, cov \, ering \, (d_i, s_j). \qquad (6)$$

Substituting (3), (4), (5), and (6) into (2), we can acquire the ultimate objective function to be minimized by the PSO.

3.3 Optimization of Detectors by PSO

The basic steps of our PSO-based NSA detectors optimization scheme are given as follows:

Step 1. Initialize the acceleration constants c_1 and c_2, inertia weight ω, maximum number of generations *MAXDT*, size of population *popsize* in the PSO. The current iteration $t=1$, and consider the position of detectors generated by the NSA $x_1=[D(1:m,1)' \, D(1:m,2)'...D(1:m,n)']$ as a particle. The (*popsize*-1) particles x_2, ..., $x_{popsize}$ randomly generated from R^{nm} constitute the initial population $X(t)$. The initial velocities v_1, v_2, ..., $v_{popsize}$ of each particle are included in the velocity matrix $V(t)$.

Step 2. Evaluate the fitness value of each particle in $X(t)$.

Step 3. Find *pbest*: Compare the evaluated fitness value of each particle with *pbest*, which is the best solution of each particle in the history. If the current value is better than *pbest*, then set the current fitness value as *pbest*, and the current particle location as the *pbest*'s location.

Step 4. Find *gbest*: If the current fitness value is better than *gbest*, which is the best solution of the whole population, then set *gbest* to be the current fitness.

Step 5. Update the velocities and locations of the particles according to (7) and (8), and obtain the next population $X(t+1)$:

$$v_i(t+1) = w*v_i(t) + c_1*r_1*(pb_i - x_i(t)) + c_2*r_2*(pg_i - x_i(t)), \qquad (7)$$

$$x_i(t+1) = x_i(t) + v_i(t+1). \qquad (8)$$

Step 6. Repeat from Step 2 to Step 5 until a given termination criterion is met, which is usually a sufficiently good fitness value or a predefined maximum number of generations *MAXDT*. Some notions used here are explained as follows: t is iteration index, ω is the inertia weight, which is to balance both the local and global search, c_1 and c_2 are the two acceleration constants, usually $c_1=c_2=2$, r_1 and r_2 are two random values in the range [0, 1], $x_i(t)$ is the current location of the particle, pb_i is the best previous location (which results in the best fitness value) of the ith particle, pg_i is the best previous location of all the particles in the population, $v_i(t)$ is the current velocity of the particle, $v_i(t+1)$ is the updated velocity of the particle, $x_i(t+1)$ is the updated location of the particle. Note that if the updated location of any particle is beyond the solution space, this particle will be bounced back.

3.4 Displacement of Detectors after PSO Optimization

There could be some overlapping between the detectors optimized by the PSO and the self set. These detectors have to be moved away from the self set in order to decrease the false alarm rate. Totally, three different solutions are considered in this paper:

1. Random displacement.
2. Move the detectors overlapping with the self set to their original locations.
3. Move the detectors overlapping with the self set using the following algorithm:

Step 1. Judge whether the ith ($i=1,...,m$) detector overlaps with the self set. If any overlapping exists, calculate *num*, the number of the self elements, which overlap with the ith detector. Otherwise, go to Step 4.

Step 2. If *num*=1, move the detector along the vector $\overrightarrow{c_d^i c_s^j}$ or the opposite direction with the distance of $\beta*dslap(c_d^i, c_s^j)$, where $dslap(c_d^i, c_s^j)$ is the staggered distance between detector d_i and self s_j, and β is a proportional factor. If *num*>1, a smaller displacement is chosen.

Step 3. Return to Step 1, and check whether the newly generated detector overlaps with the self set.

Step 4. Check whether all the detectors have been generated. Otherwise, $i=i+1$, and return to Step 1.

Our experiments show that the performance of the random displacement is not satisfactory. The second method is moderately better than the first one. However, the third solution can provide us the best performance, and therefore it is used in the simulations in Section 4.

4   Simulation Results and Analysis

4.1   Experiments with Synthesized Datasets

To demonstrate the effectiveness of the proposed NSA detector optimization approach, a simple sine time signal as shown in Fig. 1 is first used, which is given in (9). An anomaly from 65s to 85s is added to this time series, as in (10):

$$y(i) = \sin\left(\frac{\pi}{10}i\right),\tag{9}$$

$$y(i) = \frac{1+rand}{2}\sin\left(\frac{\pi}{10}i\right),\tag{10}$$

where i is the time, and *rand* is a random value in the range [0, 1].

The sine time series is normalized. A sliding overlapping window of size one is employed to produce two-dimensional samples. The first 50 points are selected as the self set. The detectors are generated by the NSA, and the PSO is next used to optimize these detectors. The following parameters are used in our simulations: window size $n=1$, sliding shift *step*=1, self radius $Rs=0.01$, number of detectors $m=30$, acceleration constants $c_1=c_2=2$, inertia weight $\omega=0.7298$, maximum number of generations $MAXDT=1000$, size of population *popsize*=20, and weight in the objective function $W=0.02$.

The relationship between the overlapping of detectors and iteration time is illustrated in Fig. 2. Obviously, the overlapping of detectors gradually decreases with the

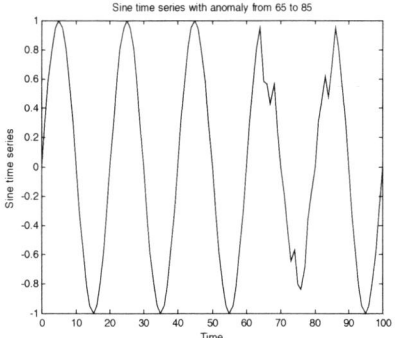
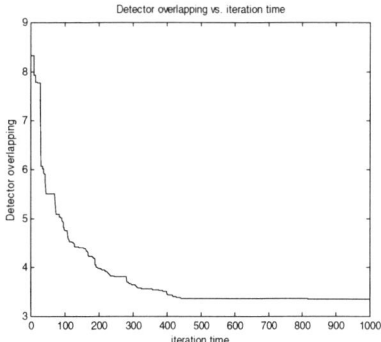

Fig. 1. Sine time series with anomaly   **Fig. 2.** Overlapping of Detectors *vs.* iteration time

growth of the iteration time. The anomaly detection rates before and after the PSO optimization are given in Table 1. It can be concluded that our PSO-based approach is capable of improving the anomaly detection rate of the regular NSA.

Table 1. Anomaly detection rates before and after optimization

	Detection rate *DR*	False alarm rate *FAR*
Before optimization	62.7%	0
After optimization	87.1%	0

4.2 Experiments with Real-World Datasets

Vibration signals of motor bearings are used here to further examine the performance of our method. Fig.3 and 4 illustrate the vibration signals of the normal bearings and those with broken balls, respectively. There are totally 20,000 samples in each of these two time series. Firstly, both time series are normalized to [0,1]. A sliding

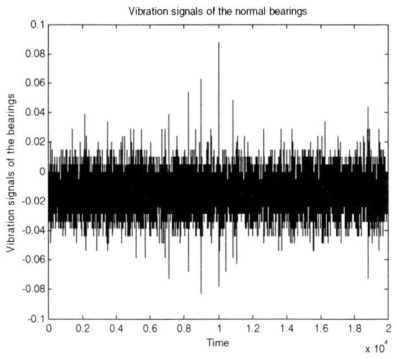
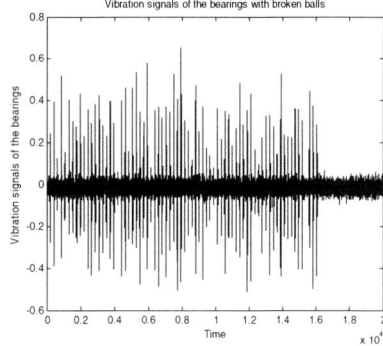

Fig. 3. Time series of healthy bearings   **Fig. 4.** Time series of faulty bearings

overlapping window of size 10 is deployed to generate ten-dimensional samples. The first 15,000 points of the normal time series are selected as the self set. The following parameters are employed in the simulations: window size $n=10$, sliding shift $step=10$, self radius $Rs=0.001$, number of detectors $m=30$, acceleration constants $c_1=c_2=2$, inertia weight $\omega=0.7298$, maximum number of generations $MAXDT=1000$, size of population $popsize=20$, and weight in the objective function $W=0.015$.

The anomaly detection results for the normal and faulty bearings time series are shown in Figs.5 and 6, respectively. It is clearly visible that the PSO-optimized detectors can achieve a satisfactory performance of detecting the fault of broken balls in the bearings. Usually, the number of detectors activated by the normal data, $Num1$, and the number of detectors activated by the abnormal data, $Num2$, are used to evaluate the efficiency of anomaly detection. Therefore, $Num1/Num2$ before and after the PSO optimization are given in Table 2, which demonstrate that our new NSA optimized by the PSO performs much better than its original version.

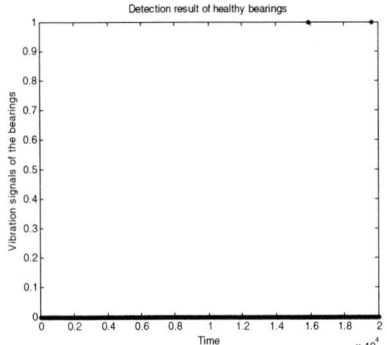 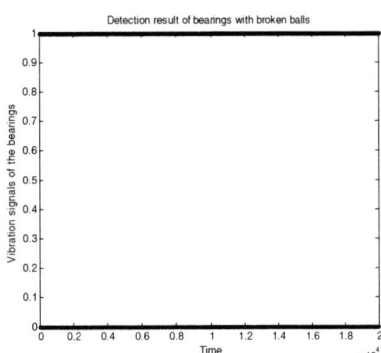

Fig. 5. Detection result of healthy bearings   **Fig. 6.** Detection result of faulty bearings

Table 2. Comparison of $Num1$ and $Num2$ before and after PSO optimization

	$Num1$	$Num2$	$Num1/Num2$
Before optimization	4	23	0.17
After optimization	2	27	0.07

5 Conclusions

As we know, achieving the maximal coverage of the non-self space with a given number of detectors is an important issue in the NSA for anomaly detection. In this paper, the PSO is employed to optimize the locations of the detectors randomly generated by the regular NSA. Numerical simulations have verified the effectiveness of our new approach. In the optimization of the NSA detectors, appropriate selection of the objective function is crucial. The calculation of the volume inside the boundaries of the unit hypercube is also difficult in case of edge hyper-spheres. Moreover, the estimation of overlapped volume of detectors can be another challenging problem, especially, when different hyper-shaped detectors are used. How to attack these problems will be our future research topics.

Acknowledgments. X. Z. Gao's research work was funded by the Academy of Finland under Grant 214144.

References

1. Hofmeyr, S., Forrest, S.: Architecture for All Artificial Immune System. Evolutionary Computation Journal, 443–473 (2000)
2. Dasgupta, D., Forrest, S.: Novelty Detection in Time Series Data Using Ideas From Immunology. In: Proc. Int. Conf. Intelligent Systems, pp. 87–92 (1996)
3. González, F.: A Study of Artificial Immune Systems Applied to Anomaly Detection. PhD. Dissertation (2003)
4. D'haeseleer, P., Forrest, S., Helman, P.: An Immunological Approach to Change Detection: Algorithms, Analysis and Implications. In: The Proceedings of IEEE Symposium on Security and Privacy, pp. 110–119 (1996)
5. Forrest, S., Perelson, A., Allen, L., Cherukuri, R.: Self-nonself Discrimination in A Computer. In: Proc. IEEE Symp. on Research in Security and Privacy, pp. 202–212 (1994)
6. González., F., Dasgupta, D., Kozma, R.: Combining Negative Selection and Classification Techniques for Anomaly Detection. In: Fogel, D.B., El-Sharkawi, M.A., Yao, X., Greenwood, G., Iba, H., Marrow, P., Shackleton, M. (eds.) Proceedings of the 2002 Congress on Evolutionary Computation CEC 2002, USA, May 2002, pp. 705–710. IEEE Press, Los Alamitos (2002)
7. Dasgupta, D., Krishna Kumar, K., Wong, D., Berry, M.: Negative Selection Algorithm for Aircraft Fault Detection. In: Proceedings of Third International Conference on Artificial Immune Systems (ICARIS 2004), pp. 1–13 (2004)
8. Ji, Z., Dasgupta, D.: Real-Valued Negative Selection using Variable-Sized Detectors. In: Proceedings of the Genetic and Evolutionary Computation Conference (GECCO), Seattle, Washington (2004)
9. Ji, Z., Dasgupta, D.: Augmented Negative Selection Algorithm with Variable-Coverage Detectors. In: Proceedings of the Congress on Evolutionary Computation (CEC), Portland, Oregon, U.S.A (2004)
10. Kennedy, J., Eberhart, R.C.: Particle Swarm Optimization. In: Proc. IEEE International Conference on Neural Networks, pp. 1942–1948. IEEE Service Center, Piscataway (1995)

Adaptation of Learning Parameters for Choquet Integral Agent Network by Using Genetic Algorithms

Santoso Handri[1] and Kazuo Nakamura[2]

[1] Graduate School of Information Science and Control Engineering
[2] Management Information and Systems Science
  Nagaoka University of Technology, 1603-1 Kamitomiokamachi, Nagaoka,
  Niigata 9402124 – Japan
  {bondry@alice.nagaokaut.ac.jp,nakamura@kjs.nagaokaut.ac.jp}

Abstract. Choquet Integral Agent Network (CHIAN) is proposed as a method realizing flexible information fusion which is constructed by using fuzzy measure and Choquet integrals. In case of multi-layered network structure, CHIAN can employ back-propagation algorithms-like concept for learning process. However, the back-propagation methods have some limitations such as trapping at local minima and network paralysis. Due to genetic algorithms (GA) mechanism, it has the characteristics of hill climbing, and thus can overcome the difficulty of trapping at local minima; consequently it might reduce network paralysis. This paper aims at proposing to tune CHIAN learning parameters, i.e., learning rate and momentum coefficient by genetic algorithms for improving CHIAN as classifier, pattern recognition, and information fusion. The results show that the network evolved GA requires fewer training cycles than the network which the learning parameters are intuitively given.

Keywords: Choquet integral agent network, genetic algorithms, back-propagation, adaptive learning parameters.

1 Introduction

In recent years, the developments of fuzzy logic, probabilistic reasoning, neural network and evolutionary algorithm motivated the researchers to explore the possibilities of building more humanlike machines using these tools. The synergisms of these tools might also improve the performance of the overall system to a great extent. It is a partnership in which each of the partners contributes a distinct methodology for addressing problems in its domain. In this perspective, the principal contributions of fuzzy logic, probabilistic reasoning, neural network and evolutionary algorithm are complementary rather than competitive [1]. As reported [2] [3] [4], the genetic algorithms have been used in conjunction with neural networks in three major ways. First, they have been used to set the weights in fixed network architectures. In related work, a genetic algorithm was also employed to set the learning rate parameters. Second, genetic algorithms have been used to learn neural network topologies, such as specifying how many hidden units a neural network should be have and how the nodes are

E. Avineri et al. (Eds.): Applications of Soft Computing, ASC 52, pp. 22–33.

connected. Third, genetic algorithms were used to select training data and to interpret the output behavior of neural networks. Thus, similar concept is tried to be employed for Choquet integral agent network (CHIAN) topology which has similar network structure with neural networks.

CHIAN is a network of many agents, each of which has integration function of the real valued multiple inputs with specific measure on the input channels and real valued single output. CHIAN employs a Choquet integral mechanism in the framework of fuzzy measure at every neuron unit instead of simple weighted sum mechanisms in neural networks. Thus the neurons may be called "agent" owing to the highly flexible information fusing ability like human intellectual processing or judgment. CHIAN has beneficial features to realize the operational means embedding existing partial and qualitative human knowledge and gray box representation [5]. To enhancing applicability of CHIAN, it has conceptually proposed for three problems, i.e., forward integration problems, system identification problems, and inverse problems [6].

The supervised learning of CHIAN is fulfilled by presenting training patterns, which map to specified features. It employs back-propagation training with gradient-descent algorithm to reduce the error along its gradient for improving the performance of the networks. However, the initial choice of the basic parameters, i.e., network topology, fuzzy measure, learning rate, and other parameters are arbitrary set for conventional problems. The selection of those parameters follows in practical use of rules of thumb, but their values are at most arguable. The method which employs back-propagation training algorithms also has limitation trapping in local minima. In this way, this paper proposes the use of genetic algorithms (GA) to tune the learning parameters of CHIAN for appropriate network structure. The objective of this paper is to optimize the learning process in pursuing an optimal performance of CHIAN.

The outline of this paper is the following. Section 2 presents the outline of CHIAN and its algorithms. This section describes the back-propagation algorithms-like concepts for multi-layered CHIAN problems. Section 3 presents the basic theory and practical use of GA in machine learning for tuning CHIAN learning parameters. Section 4 shows the experimental results of this paper. In the experiments, the proposed methods also are evaluated. Finally, discussion and conclusions are presented in Section 5.

2 CHIAN Concept

CHIAN is a soft computing approach using flexible modeling. Explanation of CHIAN is following. CHIAN is quasi-hierarchical network of units making a Choquet integral of multiple inputs in the interval [0, 1] with assigned fuzzy measures. This computational mechanism has a feature realizing much more flexible information fusion comparing to neural network. The skill based and ruled based cognitive knowledge could be embedded simultaneously and hierarchically in the mechanism.

2.1 Expression of Each Agent in CHIAN

The schematic diagram of an agent in CHIAN is shown in Fig. 1. The figure description is as follow; the j-th agent of k-th layer gets the normalized quantitative or qualitative inputs among system input data $\{u_i\}$ from the real world if $k=1$ or among the outputs of the agents $\{x_j^{k-1}\}$ as the multiple inputs $\{f(s)\}$, and they are integrated by Choquet integral mechanism which results in normalized output value $\{x_j^k\}$ as the meaningful factor. The mathematical expression is as follow:

Let $S = \{s_1, s_2, \cdots, s_m\}$ be a collection of input channels for an agent, ϕ be the empty set and let a set function $w : 2^S \rightarrow [0,1]$ be a fuzzy measure satisfying the properties;

(i)   boundedness : $w(\phi) = 0, w(1) = 1,$

(ii)   monotonicity : $A \subseteq B \subseteq S \Rightarrow w(A) \le w(B).$

Then, setting the real valued function $f : S \rightarrow [0, \infty]$ as integrated function, i.e., multiple input values, the obtained result x by Choquet integral with respect to fuzzy measure μ is defined by;

$$ı x = C(f, \mu) = (c) \int f(s) d\mu ı$$

$$= \sum_{r=1}^{m} [f(p(r)) - f(p(r-1))] \cdot w(A_{(r)}) \tag{1-a}$$

$$= \sum_{r=1}^{m} f(p(r)) \cdot [w(A_{(r)}) - w(A_{(r-1)})] \tag{1-b}$$

where $p(r)$ be a permutation of elements of S so that ı

$$f(p(0)) = 0, 0 \le f(p(1)) \le f(p(2)) \le \cdots \le f(p(m)) \le 1, \text{ı}$$

$$A_{(r)} = \{p(r), p(r+1), \cdots, p(m)\}, A(m+1) = \phi. \text{ı}$$

As known well Choquet integral has the specific properties as a quite flexible aggregation. That is, it includes Lebesgue integral which corresponds to traditional weighted sum in the case of finite input sets, and simultaneously includes logical aggregation of fuzzy features in terms of multiple qualitative items.

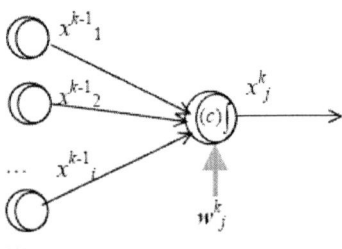

Fig. 1. Schematic diagram of an agent in CHIAN

2.2 CHIAN-Based Back-Propagation Algorithms

Further, the multi-layered network is illustrated to derive the proposed algorithm for supervised CHIAN structure. Suppose the CHIAN with a single hidden layer is shown in Fig. 2. In this structure, the CHIAN agent in each layer of the network is connected to every other agent in the adjacent layer. In other words, the input layer projects onto a hidden layer of CHIAN and the output of the hidden layer projects onto the output layer. Let x_j and x_k be the state of j-th agent of hidden layer and k-th agent of output layer, respectively. The fuzzy measure subsets of the hidden layer and output layer are denoted by v_j and w_j, respectively. The activation function of the hidden layer and the output layer is given by $f(.)$ and $g(.)$, respectively. Then the processing for the forward integration is expressed by;

$$z_j = (c)\int u \cdot dv_j = \sum_r \left[u_{p(r)} - u_{p(r-1)}\right] \cdot v_{j,r} = \sum_r u_{p(r)} \left[v_{j,r} - v_{j,r+1}\right] \tag{2}$$

$$x_j = f(z_j) = 1/\left(1 + e^{-z_j}\right) \tag{3}$$

$$z_k = (c)\int x_j \cdot dw_k = \sum_t \left[x_{j(t)} - x_{j(t-1)}\right] \cdot w_{k,t} = \sum_t x_{j(t)} \left[w_{k,t} - w_{k,t+1}\right] \tag{4}$$

$$x_k = f(z_k) = 1/\left(1 + e^{-z_k}\right) \tag{5}$$

The back-propagation algorithm is then employed in the learning process for identifying CHIAN. The purpose of learning process is to determine the best set of fuzzy measure for the network in such a way that the discrepancy between the desired and actual output is minimized. The given desired system output is denoted by d_k. One measure that is commonly used as discrepancy is the sum of squared error by;

$$E = \frac{1}{2}\sum_k (d_k - x_k)^2 \tag{6}$$

Then, the fuzzy measure of hidden layer is updated using the following equation based on gradient descent algorithms;

$$\Delta v = -\eta \frac{\partial E}{\partial v}$$

$$\frac{\partial E}{\partial v} = \frac{\partial E}{\partial x_k} \frac{\partial x_k}{\partial z_k} \frac{\partial z_k}{\partial x_j} \frac{\partial x_j}{\partial z_j} \frac{\partial z_j}{\partial v}$$

$$E = \frac{1}{2}\sum (d_k - x_k)^2 \tag{7}$$

$$\frac{\partial E}{\partial z_k} = -(d_k - x_k)\frac{\partial x_k}{\partial z_k} = -(d_k - x_k)\cdot x_k \cdot (1 - x_k)$$

$$\delta_k = x_k \cdot (1 - x_k)(d_k - x_k)$$

$$\frac{\partial E}{\partial v} = \sum_j \left(-\sum_k \delta_k \frac{\partial z_k}{\partial x_j} \right) \frac{\partial x_j}{\partial z_j} \frac{\partial z_j}{\partial v}$$

$$\frac{\partial E}{\partial v} = -\sum_j x_j \cdot (1 - x_j) \cdot \left(\sum_k \delta_k \left(w_{k,t} - w_{k,t+1} \right) \right) \cdot \left[u_{p(r)} - u_{p(r-1)} \right]$$

$$\delta_j = x_j \cdot (1 - x_j) \cdot \left(\sum_k \delta_k \left(w_{k,t} - w_{k,t+1} \right) \right)$$ (8)

$$\frac{\partial E}{\partial v} = -\sum_j \delta_j \cdot \left[u_{p(r)} - u_{p(r-1)} \right]$$

Thus the recursive revision of the estimated fuzzy measure of hidden layer is performed by;

$$\Delta v = -\eta \frac{\partial E}{\partial v} = \eta \cdot \sum_j \delta_j \cdot \left[u_{p(r)} - u_{p(r-1)} \right] + \mu \cdot \Delta v_{t-1}$$ (9)

where η and μ are learning rate and momentum coefficient parameter, respectively.

Using the expression derived above, the general representation of multi-layered CHIAN for updating fuzzy measure of each agent can be achieved by the learning algorithms described below;

In term of the *j-th* agent in *k-th* layer the state is obtained by Choquet integral.

$$z_j^1 = \sum_r \left[u_{p_1(r)} - u_{p_1(r-1)} \right] \cdot w_{j,r}^1$$ (10)

$$z_j^k = \sum_r \left[x_{p_1(r)}^{k-1} - x_{p_1(r-1)}^{k-1} \right] \cdot w_{j,r}^k \quad for \, k = 2,3,\cdots,n$$ (11)

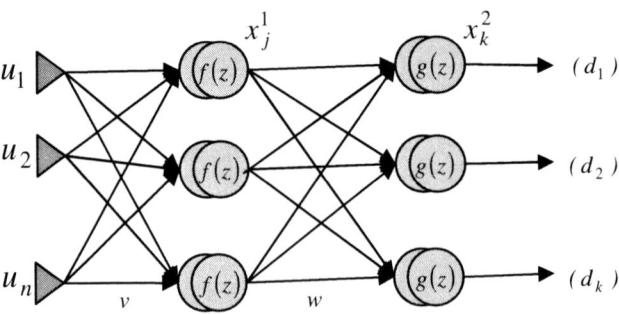

Fig. 2. A CHIAN with one hidden layer and output layer

In this structure, it is assumed that each agent have logistic sigmoid activation function. A useful feature of this function is that its derivative can be expressed in a particularly simple form;

$$x'(z) = x(z) \cdot (1 - x(z)) \tag{12}$$

The sum of the squared error function is defined by;

$$E = \frac{1}{2} \sum_j \left(d_j - x_j^n \right)^2 \tag{13}$$

According to the general expression of the network for back-propagation, together with (12) (13), we obtain the following results. For the output unit, the $\delta's$ are given by;

$$\delta_j^n = x_j^n \left(1 - x_j^n \right) \left(d_j - x_j^n \right) \tag{14}$$

while for agents in the hidden layer the $\delta's$ are found using:

$$\delta_j^k = x_j^k \left(1 - x_j^k \right) \left(\sum_l \delta_l^{k+1} \left(w_{l,p_{k+1}^{-1}(j)}^{k+1} - w_{l,q_{k+1}^{-1}(j)+1}^{k+1} \right) \right) , k = 1,2,\cdots,n-1 \tag{15}$$

Then, the correction term of fuzzy measure w to be updated corresponding to the given training data is obtained by;

$$\Delta w_t^1 = -\eta \frac{\partial E}{\partial w^1} = \eta \cdot \sum_l \delta_l^1 \cdot \left[u_{p(r)} - u_{p(r-1)} \right] + \mu \cdot \Delta w_{t-1}^1 \tag{16}$$

$$\Delta w_t^k = \eta \cdot \sum_l \delta_l^k \cdot \left[x_{l,p(r)}^k - x_{l,p(r-1)}^k \right] + \mu \cdot \Delta w_{t-1}^k \; for \, k = 2,3,\cdots,n \tag{17}$$

After computing the backward propagated error δ_j^k at the k-th layer consecutively, the correction term for each estimated fuzzy measure w is obtained at every recursive procedure.

3   Optimization of Learning Parameters for CHIAN Identification by Genetic Algorithms

It is often overlooked that the identification performance of a CHIAN on a certain problem depends on the network architecture used and the actual knowledge representation (i.e., values of the fuzzy measure and learning rate) within that specific architecture. It can be said that the identification performance of a CHIAN depends on three factors: the problem for which the network is going to be used, the network structure and the set of fuzzy measure. The performance of a CHIAN is typically measured by the cumulative error of the network on some training data with known target outputs, but can include computational speed and complexity as well. This performance can be defined by an abstract quality function;

$$E = E \, (Tr, \, S, \, W, \, L) \tag{18}$$

where,

      E = the type of quality function.
      Tr = the training data (i.e. the target input/output data set).

S = the structure of the network.
W = the set of fuzzy measure.
L= Learning parameters

The objective is to optimize the quality function E to gain an optimal performance of CHIAN. This really holds for any type of CHIAN topology. This study only deals with CHIAN identification problems that can be trained with some type of supervised learning algorithm. An example of such a quality function E that is commonly used is the mean cumulative squared error on the training data set consisting of several input / output pattern.

$$mse = \frac{1}{2P} \sum_p \left(T_p - O_p\right)^2 \tag{19}$$

where,

T = the target vector.
O = the CHIAN output vector.
P = the number of test pattern.

Traditionally the structure of a CHIAN, S, is set by the user a priori. The type of structure used may be based on some knowledge of the problem domain but commonly a sufficient network structure is built based on the experiment by testing several networks model. In many cases the structure used will be a fully connected network and the user might try different numbers of hidden agents to see how well the resulting structures will fit the task. This network structure is then trained with some learning algorithm to gain an appropriate set of fuzzy measure, W. The emphasis on optimizing the quality of the network, E, is very often based on the ability of the learning algorithm to generate an optimal set of fuzzy measure, while the structure S is taken for granted or chosen from a limited domain.

The automatic generation of a network structure is a useful concept, as, in many applications, the optimal structure is not known a priori. The conventional algorithms have several drawbacks. They are usually restricted to a certain subset of the network topologies and as with hill climbing methods they often stuck at local optima and may therefore not reach the optimal solution. These limitations can be overcome using genetic algorithms as approach to the generation of CHIAN structures.

Genetic algorithms (GA) are iterative search algorithms based on an analogy with the process of natural selection (Darwinism) and evolutionary genetics. The main goal is to search for a solution, which optimizes a user-defined function called fitness function. To perform this task, it maintains a population for each generation. The population consists of individuals or chromosomes, and each chromosome is contained the genes. The chromosome is generated by randomly following a uniform, Gaussian distribution over search space or initially given intuitively. The gene type of chromosome can be binary strings or real value. Then, a new population is formed by selecting the more fit chromosomes. Some members of the new population undergo transformation by means of genetic operators to form new solutions. After some generations, it hoped that the best chromosome represent a near-optimal solution. There are three operators in GA, i.e., selection, crossover and mutation [7]. The selection is the process selecting the chromosome in a population for further genetic operations.

Each chromosome in a population is assigned a value of fitness. The fitness values are used to assign a probability value to each chromosome. The chromosome with a larger fitness value has a larger probability of selection. The crossover operation combined the features of two parent chromosomes to form two similar offspring by swapping corresponding segments of the parents. The parameters defining the crossover operation are the probability of crossover and the crossover position. Mutation is a process of occasional alternation of some gene values in a chromosome by a random change with a probability less than the mutation rate.

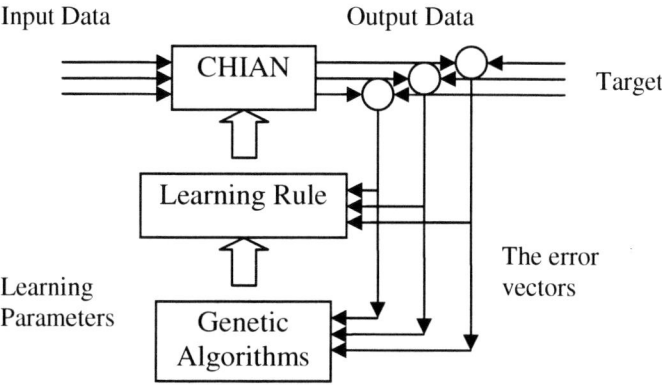

Fig. 3. The framework of learning parameters adaptation for CHIAN identification by using genetic algorithms

In this study, GA is used to estimate learning rate (η) and momentum coefficient (μ) parameters of CHIAN by minimize the mean square error. Following the training process as explained in Fig. 3, it starts with constructing the CHIAN by setting the total layers and agents of each layer. The fuzzy measures are then set randomly. The chromosomes are constructed with: η and μ as the genes. The fitness of chromosomes is measured by the mean square error signals (target-output) at the output layered. Then the GA operations, i.e., selection, crossover and mutation are generated until the maximum iteration is reach or scores of fitness function are minimal below than a certain threshold. After a number of genetic evolutions, GA determines the nearly optimal values of the learning parameters. The overall optimized algorithm is listed below:

```
Algorithm 1: Tuning CHIAN learning parameters by ge-
netic algorithms

Step 1: Initialize CHIAN structure, Initialize fuzzy
measures, w, randomly. Set values of maximum genera-
tion, threshold, and initial population.
Step 2:
for generation = 1: maximum generation
```

```
for pop = 1: population Size
      update the fuzzy measures using (16) (17)
      count score of fitness function using (19)
   end
   Perform selection, crossover, and mutation
End
optimal η and μ are found.
```

4 Experiment and Results

The application of the GA in CHIAN structure will be illustrated in this section. The identification problem, i.e., three inputs, one hidden layer and an output layer with three outputs are constructed as shown in Fig. 2. The outputs training data are generated by forward CHIAN mechanism with given input and randomly given fuzzy measures. Then the CHIAN is trained by the given training data and the initial fuzzy measures. In the first stage, GA is employed to estimate learning rate (η) and momentum coefficient (μ) parameters. The second stage, the estimated learning parameters is used together with the training data and the initial fuzzy measures for the trained network. The comparative results of the proposed and conventional methods for obtaining the trained networks are shown in Fig. 4. It shows that the conversion error of adaptation learning parameters of CHIAN for each output channel (error 1, error 2 and error 3) is faster convergence than the given learning parameters. Thus,

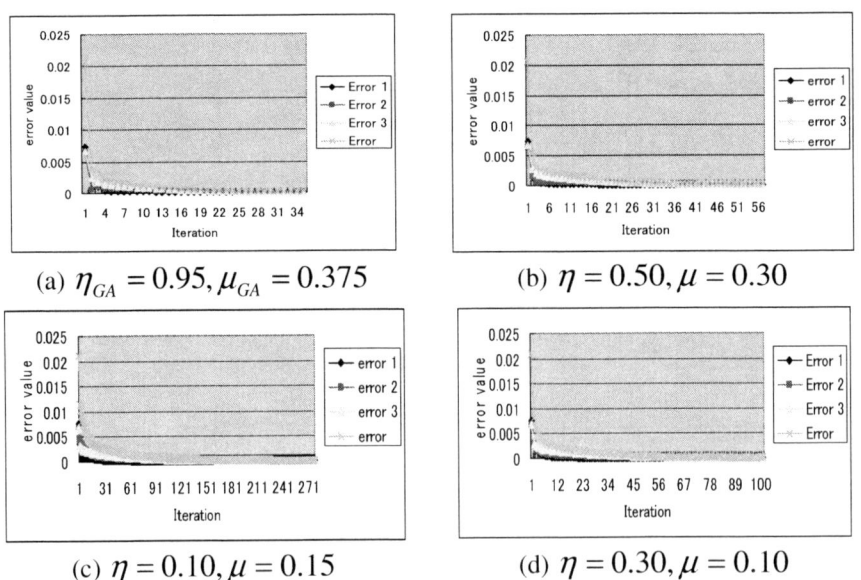

(a) $\eta_{GA} = 0.95, \mu_{GA} = 0.375$

(b) $\eta = 0.50, \mu = 0.30$

(c) $\eta = 0.10, \mu = 0.15$

(d) $\eta = 0.30, \mu = 0.10$

Fig. 4. Error comparison between adaptive CHIAN approached by GA (a) and conventional one for two layers with three agents for each (b)-(d)

the network evolved by the GA method (Fig. 4(a)) require fewer training cycles than the network which the learning parameters are intuitively given (Fig. 4(b) – (d)).

The second problem is to classify the Iris dataset. The Iris flower dataset is a popular multivariate dataset that was introduced by R.A. Fisher as an example for discriminant analysis [8]. The dataset consists of 50 samples from each of three species of Iris flowers, i.e., Iris setosa, Iris versicolor and Iris virginica. Four features were measured from each sample; they are the length and the width of sepal and petal. The iris data distribution is shown in Fig. 5. It shows that the Iris setosa is linearly separable from the other two; the latter are not linearly separable from each other, then the Iris dataset is assumed as non-linear input vector. Feature extraction is then employed to map the input vector of observation onto a new feature description which is more suitable for given task. In this study, generalized discriminant analysis [9] is used to extract the input vector of Iris dataset. The obtained features are then normalized as inputs for CHIAN and the network is constructed as illustrated in Fig. 6. Fig 7 shows the comparison of convergent error between estimated and given learning parameters of Iris problem. It shows that the result of estimated learning parameters were achieved faster and not trapped in local minima. On average, the classification accuracy was achieved in 85.7% and 98.2% with given and estimated learning parameters, respectively.

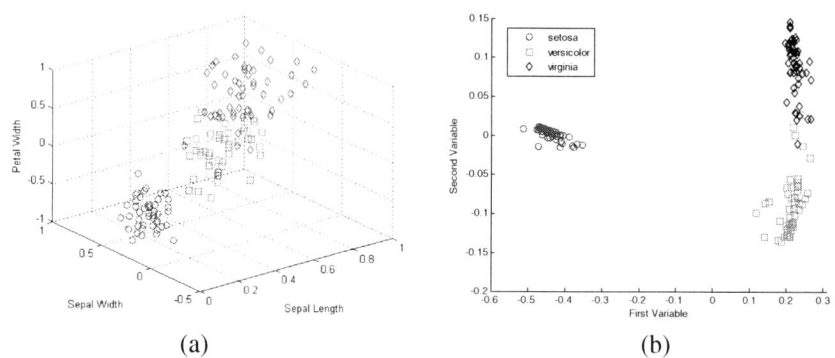

(a)   (b)

Fig. 5. Data distribution of Iris data. (a). Original data. (b). The feature extraction results.

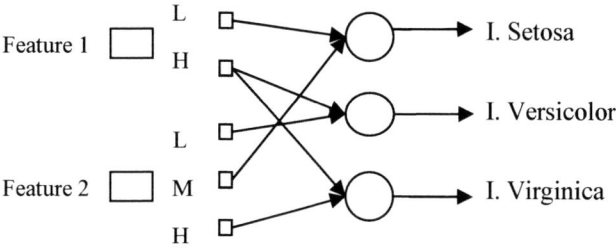

Fig. 6. The CHIAN structure for Iris dataset classification

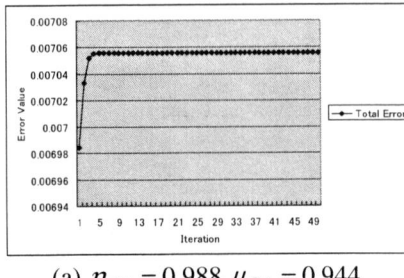

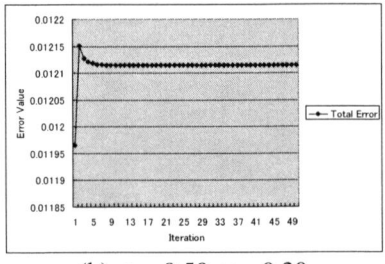

(a) $\eta_{GA} = 0.988, \mu_{GA} = 0.944$ (b) $\eta = 0.50, \mu = 0.30$

Fig. 7. Error comparison between adaptive CHIAN approached by GA (a) and conventional one (b) for Iris problem

In this study, a simple GA is used to estimate learning parameters of CHIAN. Population size and generation are set to 30 and 50, respectively. The selection process used is Roulette-wheel selection. Probabilities of the crossover and a mutation rate are set to 0.8 and 0.2, respectively.

5   Conclusions

This study has accomplished adaptation learning parameters of Choquet integral agent network (CHIAN) using genetic algorithms (GA). GA has advantages rather than the network use converging optimization method, in case, the given problem has many local minimum solutions. This work has demonstrated the real-world application of CHIAN together with GA. It shows that the proposed method was effectively optimizing learning parameters of CHIAN for several problems. This result also will help the users who are interested in employment of CHIAN as classifier, pattern recognition or information fusion. However, the result of the iris flower problem showed that the errors converged still remaining, it can be caused by the noisy observed data or the CHIAN network structure which was not appropriate constructed. It notices that employing GA increased the computational cost in the learning process of CHIAN.

References

1 The Berkeley Initiative in Soft Computing website,
   http://www-bisc.cs.berkeley.edu/BISCProgram/History.htm
2 Takahashi, H., Agui, T., Nagahashi, H.: Designing Adaptive Neural Network Architectures and their Learning Parameters using Genetic Algorithms. In: Ruck, D.W. (ed.) Science of Artificial Neural Networks II. Proceeding of SPIE, vol. 1966, pp. 208–215 (1993)
3 Konar, A.: Behavioral Synergism of Soft Computing Tools. In: Computational Intelligent: Principles, Techniques and Applications. Springer, Heidelberg (2005)
4 Konar, A.: Genetic Algorithms. In: Artificial Intelligence and Soft Computing: Behavioral and Cognitive Modeling of the Human Brain. CRC, Boca Raton (1999)
5 Nakamura, K.: A Scheme for information fusion by Choquet Integral Agent Networks. In: Eighth IFSA Congress, pp. 954–958 (1999)

6 Nakamura, K.: Towards Inverse Problems of Choquet Integral Agent Networks as Information Fusion Mechanisms. In: Proceeding of the 6th International Symposium on Advanced Intelligent Systems, pp. 675–678 (2005)

7 Goldberg, D.E.: Genetic Algorithms in Search, Optimization, and Machine Learning. Addison-Wesley, Reading (1989)

8 Asuncion, A., Newman, D.J.: UCI Machine Learning Repository, Irvine, CA: University of California, Department of Information and Computer Science (2007),
   `http://www.ics.uci.edu/~mlearn/MLRepository.html`

9 Baudat, G., Anour, F.: Generalized Discriminant Analysis using a Kernel Approach 12(10), 2385–2404 (2000)

Comparison of Effective Assessment Functions for Optimized Sensor System Design

Kuncup Iswandy and Andreas König

Institute of Integrated Sensor Systems, University of Kaiserslautern, D-67663 Kaiserslautern, Germany
{kuncup,koenig}@eit.uni-kl.de

Summary. Currently, the design of the signal processing and recognition architecture for intelligent sensor systems is still a tedious and labor-intensive task. Optimization techniques, e.g., from gradient descent, stochastic search or evolutionary computation are available to accelerate and automate this procedure. However, appropriate assessment or cost functions are required. This paper presents effective state-of-the-art techniques and compares them to two novel, salient modifications. These are a normalized compactness measure, which has salient properties as it is non-parametric, easy-to-use, fine-grained, competitive performing, and the sum-volumetric k-NN classifier. The aims of this paper are to compare the computation complexities, discriminant properties or capabilities, and sensitivity to control parameter. The methods will be described and compared for the task of dimension reduction by feature selection. Achieved results underpin their saliency for general optimized sensor system design.

Keywords: assessment function, automated design, dimensionality reduction, sensor system optimization.

1 Introduction

The design of intelligent sensors systems, in addition to appropriate hardware choice, requires the definition of a signal processing and recognition architecture. A simplified block diagram of a sensor system is given in Fig. 1. The predominantly manual design process goes through the coarse steps of sensor selection and scene optimization, choice of signal and feature processing, dimensionality reduction, and classification. The individual steps not necessarily are limited to single methods in the blocks, but more complex graph structures can be underlying here. For instance, feature level fusion is combining or accumulating between two or more methods of feature computation techniques from single sensor or multisensor to obtain more accurately information about the object. Many information provided by feature computation methods give the possibilities for selecting of the optimal features by feature selection or dimension reduction methods [1], which can result in the high detection accuracy. The task of method selecting, combining, and setting parameters is carried out expert-driven as a tedious, time and labor consuming task with potentially suboptimal outcome.

Activities to automate the configuration and the whole design process can be found since the early nineties in the field of image recognition design. More recent approaches

E. Avineri et al. (Eds.): Applications of Soft Computing, ASC 52, pp. 34–42.
springerlink.com   © Springer-Verlag Berlin Heidelberg 2009

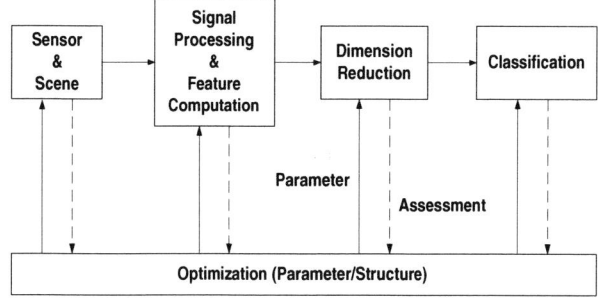

Fig. 1. The general block diagram of an intelligent sensor system

extend the concept of automation system design to the general field of intelligent sensor systems, e.g., gas sensors [2]. Optimization methods, e.g., gradient descent, stochastic search, or evolutionary computation techniques, are widely available. However, for efficient application, the block and system performance must be assessed by robust, fast, easy-to-use, and non-parametric functions. Commonly, two approaches are distinguished. The wrapper approach assesses every block modification by the overall classification output. The filter approach uses local assessment functions at each individual block. Depending on the classifier type, e.g., back-propagation networks, wrapper approach can be tedious or even infeasible to use in optimization loops. This paper presents and compares efficient and convenient assessment measures for the case of filter approach in feature selection.

2   Assessment Measures

Commonly, the voting k-NN classifier [10] is applied as assessment function [3]. In addition, other nearest neighbor based techniques, e.g., separability and overlap [4] as well as a compactness measure [4] are available. Also, the volumetric k-NN [6] is used and our sum-volumetric k-NN classifier, employs the sum of all the k nearest neighbors per class, ω_L, $L = \{1, 2, ..., n\}$, for its decision.

The nonparametric overlap measure q_o, which was inspired by the nearest neighbor concepts, provides a very fine-grained value range [8]. This normalized measure gives values close to one for non-overlapping class regions and decreases towards zero proportional to increasingly overlapping of class regions. The basic idea of overlap measure is illustrated in Fig. 2(a). The overlap measure is computed by:

$$q_o = \frac{1}{L} \sum_{c=1}^{L} \frac{1}{N_c} \sum_{j=1}^{N_c} \frac{\sum_{i=1}^{k} q_{NN_{ji}} + \sum_{i=1}^{k} n_i}{2 \sum_{i=1}^{k} n_i} \tag{1}$$

with

$$n_i = 1 - \frac{d_{NN_{ji}}}{d_{NN_{jk}}} \tag{2}$$

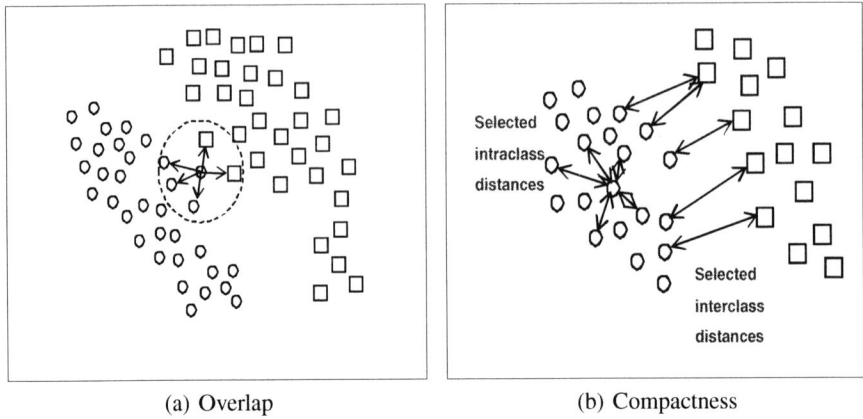

(a) Overlap   (b) Compactness

Fig. 2. Illustration of class overlap and compactness assessment

and

$$q_{NN_{ji}} = \begin{cases} n_i & \omega_j = \omega_i \\ -n_i & \omega_j \neq \omega_i. \end{cases} \tag{3}$$

Here, n_i denotes the weighting factor for the position of the ith nearest neighbor NN_{ji}, $d_{NN_{ji}}$ denotes the Euclidean distance between z_j (feature vector) and NN_{ji}, $d_{NN_{jk}}$ denotes the distance between z_j and most distant nearest neighbor NN_{jk}, $q_{NN_{jk}}$ denotes the measure contribution of z_j with regard to NN_{ji}, and ω denotes the class affiliation of z and L is the number of classes. Typically, number of nearest neighbors k well suited for computation of this quality measure is set from 5 to 10. In our experiments, we use five nearest neighbors.

The compactness measure of [8], inspired by, e.g., linear and non-linear discriminant analysis [10], still suffers from lack of normalization. Here, we extend to the normalized measure by employing normalized distances (Euclidean distance), as follows:

$$d'_{z_{ij}} = \frac{d_{max} - d_{z_{ij}}}{d_{max} - d_{min}}. \tag{4}$$

where $d_{z_{ij}}$ denotes distance between z_i and z_j. The class compactness and separation are computed by:

$$q_{Intra} = \frac{1}{L} \sum_{l=1}^{L} \frac{2}{N_l(N_l-1)} \sum_{i=1}^{N-1} \sum_{j=i+1}^{N} \delta(\omega_i, \omega_j)\delta(\omega_i, l)d'_{z_{ij}}, \tag{5}$$

$$q_{Inter} = \frac{1}{NB} \sum_{i=1}^{N-1} \sum_{j=i+1}^{N} (1 - \delta(\omega_i, \omega_j))d'_{z_{ij}}, \tag{6}$$

$$NB = \sum_{i=1}^{N-1} \sum_{j=i+1}^{N} (1 - \delta(\omega_i, \omega_j)), \tag{7}$$

where N_l is number of patterns in the L-class, N is number of all patterns, $\delta(\omega_i, \omega_j) = 1$ for $\omega_i = \omega_j$, and 0 otherwise. Figure 2(b) illustrates the principal idea of intraclass and interclass distance computation. The improved compactness assessment is an accumulation for class compactness and separation, where it can be considered as a multi-objective issue. The compactness measure is computed as following:

$$q_c = \frac{(1 - q_{Intra}) + q_{Inter}}{2}, \tag{8}$$

where this normalize compactness measurement gives a value in the range [0..1].

3   Binary Particle Swarm Optimization

Particle swarm optimization (PSO) has been used widely to solve combination optimization problems. In this paper, particle swarm algorithms are applied to obtain the optimal feature space with optimizing the assessment measures. PSO is a non-linear method affiliated to evolutionary computation techniques. Particle swarms explore the search space through a population of particles or solution x_i, which adapted by returning to previously successful regions [12]. The particles then fly over the state space, remembering the best solution encountered. The fitness function is determined by an application-specific objective function. During each iteration, the velocity of each particle is adjusted based on its momentum and influence of the best solutions encountered by itself and its neighbors. The particles then move to a new position, and the process is repeated for a prescribed number of iterations.

The original PSO technique is designed for the real-value problems. Feature selection, which is a special case of a more general technique known as feature weighting, only uses binary values to represent whether one feature is selected or not. Kennedy and Eberhart [13] have proposed Binary PSO (BPSO), where the velocity is used as a probability to determine whether the components of x_i will be in one state or zero (a binary). Previous success works related to BPSO for feature selection can be found in [5, 14, 15].

In the BPSO implementation, the trajectory of each particle is governed by the equations:

$$\mathbf{v}_i(t+1) = \omega(t) \cdot \mathbf{v}_i(t) + c_1 \cdot rand() \cdot (\mathbf{p}_i - \mathbf{x}_i(t)) + c_2 \cdot rand() \cdot (\mathbf{p}_g - \mathbf{x}_i(t)), \tag{9}$$

$$s(\mathbf{v}_i) = 1/(1 + \exp(-\mathbf{v}_i)), \tag{10}$$

and

$$\mathbf{x}_i(t+1) = \begin{cases} 1 & s(\mathbf{v}_i) > rand() \\ 0 & s(\mathbf{v}_i) \leq rand(). \end{cases} \tag{11}$$

where $\mathbf{x}_i = (x_{i1}, x_{i2}, ..., x_{iD})$ and $\mathbf{v}_i$ are the current vector position and velocity of the i-th particle, $\mathbf{p}_i$ is the position of the best state visited by the i-th particle, $\mathbf{p}_g$ is the particle with the best fitness in the neighborhood of i, and t is the iteration number. The parameter c_1 and c_2 are called the cognitive and social learning rates. The parameter ω is an inertia weight, which used to dampen the velocity during the course of the simulation, and allow the swarm to converge with greater precision. The inertia weight will decrease through the time and compute as follows

$$\omega(t+1) = \omega(t) - \frac{\omega_{start} - \omega_{end}}{0.95 \times t_{max}}, \qquad (12)$$

where the ω_{start} and ω_{end} denotes the starting and ending values of the inertia weight respectively. The parameter values of ω_{start}, ω_{end}, c_1, and c_2 used in all of our experiments were set as 1, 0.7, 2, and 2 respectively.

To determine the best particle of population in each generation, the cost function will be computed as follows

$$f_l(t) = w_q \cdot q_l + w_C \cdot C_l \qquad (13)$$

and

$$C_l(t) = 1 - \frac{Sn_l}{St}, \qquad (14)$$

where q_l denotes the assessment function (see section 2) of member l, Sn_l is number of selected features of member l, and St is total number of features. The weight parameters w_q and w_C are set as 0.99 and 0.01, where $\sum w = 1$.

4   Experiments and Results

To compare the assessment methods introduced in the previous section, four data sets were investigated. **Wine** data stems from chemical analysis of wines with 3 classes, 13 features and 59, 71, and 48 patterns per class [11]. **Rock** data stems of 134 Portuguese rocks with 18 features of oxide composition and measurements [9]. In the experiment here, we use only 110 sample rocks with regarding to three different types of rocks (granite, marble and limestone). There are 31, 51, and 28 samples of each of the three classes, respectively. **Eye-tracking** image data originated from several pictures and scenes of persons in frontal view position [7]. From these grey value images, 17 by 17 pixel regions for eye and non-eye patterns were extracted. From these image blocks 12 dimensional Gabor jets, 13 dimensional extended of the local autocorrelation (ELAC), and 33 dimensional local-orientation-coding (LOC) were computed. So eye-tracking image data consist of 58 dimensional feature data with two classes for eye (28 samples) and non-eye patterns (105 samples). The fourth benchmark data set comes from the domain of **Mechatronics**, where the compressor of a turbine jet engine was monitored with regard to its operating range [6]. The 24-dimensional data, which has also high intrinsic dimension, was generated from Fourier spectra obtained from the compressor. Four operating regions were defined as classes. In summary, 1775 samples were drawn from a compressor setup and recorded as a data set including attribute values. The objective of the underlying research was the development of a stall margin indicator for optimum jet engine operation. In our experiment we used only 700 samples, which consist of 100, 200, 350 and 50 samples of each of the four classes, respectively.

For the experiments, the holdout approach was used, dividing the data into two sets, i.e., training and testing sets. Wine data was divided into 89 samples for both training

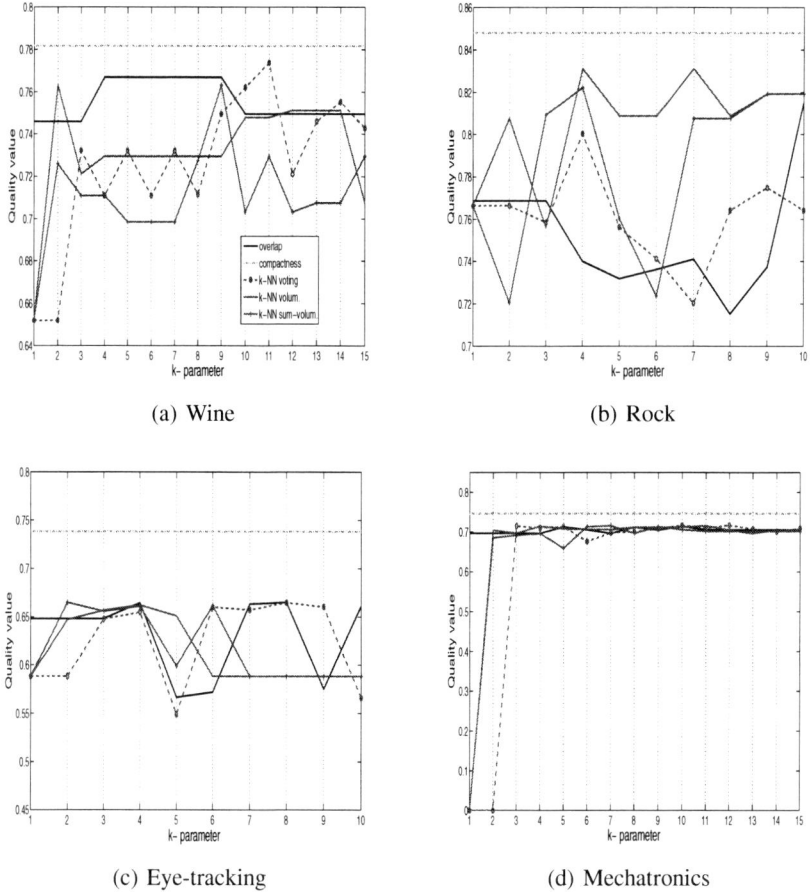

(a) Wine

(b) Rock

(c) Eye-tracking

(d) Mechatronics

Fig. 3. The parameter sensitivity and quality values of parametric methods (i.e., overlap, k-NN voting, volumetric and sum-volumetric) and nonparametric (compactness) after the learning process

and testing set. Rock data was 55 samples for both training and testing set. Eye-tracking data was split into 72 and 61 samples for training and testing set, respectively. Mechatronics data was divided into 350 samples per each subset. The training set was applied for Feature Selection (FS), employing stochastic search (i.e., BPSO), and each of the regarded assessment functions. The population of BPSO is set to 20 particles for each run.

In our experiments, we investigated the effect of parameter k, the neighborhood size in each of the assessment functions, for feature selection optimization. This was accomplished by varying k between 1 and 10 for Rock and Eye-tracking data, since their data size is small. Whereas for Wine and Mechatronics data, the neighborhood of k parameter was varied between 1 and 15. Figure 3 shows the quality of selected features achieved by each of the assessment functions for the four data sets. We used the normalized compactness measurement as the quality measurement. This gave us information about the quality of selected features with regard to the compactness and separation between samples of intra- and interclass affiliation. The feature selection using normalized compactness assessment function showed better quality values than achieved by the parametric assessment functions.

In these experiments, we compared the classification accuracies of each of the assessment functions as shown in Fig. 4, where the Wine, Rock, Eye-tracking, and Mechatronics data results are given for classification with the test set using voting 5-NN classifier. The all parametric methods (overlap measurement and all three different types of nearest neighbor assessment functions) showed a tendency of the classification increase, when the neighborhood size k is incremented. Except in experiments using Eye-tracking data set, all three k-NN assessment functions were unpredictable. Varying the k parameter does not affect the classification results of compactness measures. The number of selected features by each of the regarded assessment functions were summarized in Table 1, where all parametric methods have sensitivity to varied neighborhood size (k parameter).

Table 1. The comparison of the number of selected features after applying feature selection between parametric methods (i.e., overlap, k-NN voting, volumetric and sum-volumetric) and nonparametric (compactness)

Data set	Assessment	No. of Selected Features		
		Min.	Max.	Ave.
Wine	overlap	6	7	6.4
	compactness	5	5	5
	k-NN voting	2	6	4.2
	k-NN volum.	2	6	4.3
	k-NN sum-volum.	2	4	3.5
Rock	overlap	2	8	6.6
	compactness	5	5	5
	k-NN voting	2	5	3.4
	k-NN volum.	2	3	2.1
	k-NN sum-volum.	2	7	3.4
Eye-Tracking	overlap	14	24	18.6
	compactness	24	24	24
	k-NN voting	5	13	8.2
	k-NN volum.	5	9	6.3
	k-NN sum-volum.	5	14	7.7
Mechatronics	overlap	5	8	6
	compactness	12	12	12
	k-NN voting	1	6	4.2
	k-NN volum.	1	7	5.4
	k-NN sum-volum.	1	5	3.7

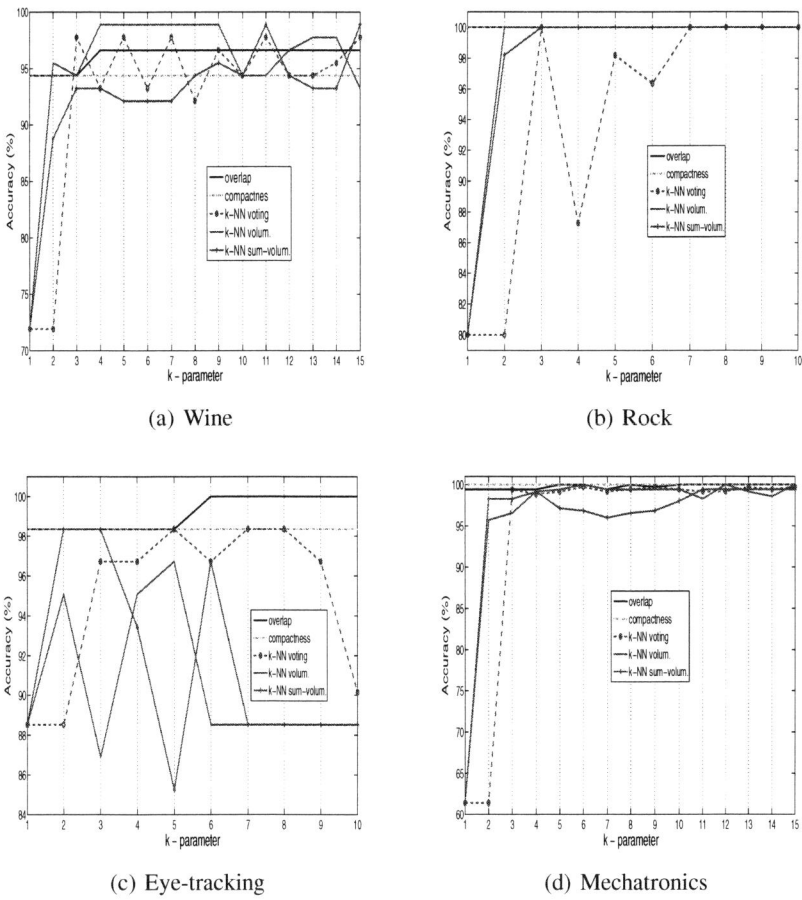

(a) Wine   (b) Rock

(c) Eye-tracking   (d) Mechatronics

Fig. 4. The comparison of recognition accuracy results after applying feature selection between parametric methods (i.e., overlap, k-NN voting, volumetric and sum-volumetric) and nonparametric (compactness) using 5-NN voting classifier

5   Conclusion

The obtained results of the assessment function comparison show, that the compactness is easy-to use and convenient, i.e., free of parameter settings. The sum-volumetric k-NN proved also to be very effective. These measures are individually applicable for intelligent sensor system design optimization with regard to recognition tasks, where the parameter selection and structure combination, e.g., sensor fusion, feature fusion, and decision fusion, can be determined and optimized. In future work, the application to automated feature computation optimization will be regarded [4]. As the measures pursue different goals in feature space, combination in a multi-objective optimization approach seems very attractive and will be pursued, including cost issues as further sub goals. The inclusion of support vector machines (SVM) will be considered in our future work for automated feature computation optimization and classification.

References

1. Guyon, I., Elisseeff, A.: An Introduction to Variable and Feature Selection. J. of Machine Learning Research 3, 1157–1182 (2003)
2. Gutierrez-Osuna, R., Nagle, H.T.: A method for evaluating data- preprocessing techniques for odor classification with an array of gas sensors. IEEE Trans. on Systems, Man and Cybernetics B 29(5), 626–632 (1999)
3. Raymer, M.L., Punch, W.F., Goodman, E.D., Kuhn, L.A., Jain, A.K.: Dimensionality Reduction Using Genetic Algorithms. IEEE Trans. on Evolutionary Computation 4(2), 164–171 (2000)
4. Iswandy, K., König, A., Fricke, T., Baumbach, M., Schuetze, A.: Towards Automated Configuration of Multi-Sensor Systems Using Evolutionary Computation - A Method and a Case Study. J. Computational and Theoretical Nanoscience 2(4), 574–582 (2005)
5. Iswandy, K., König, A.: Feature selection with acquisition cost for optimizing sensor system design. Adv. Radio Sci. 4, 135–141 (2006)
6. König, A.: Interactive visualization and analysis of hierarchical neural projections for data mining. IEEE Trans. on Neural Networks 11(3), 615–624 (2000)
7. König, A., Mayr, C., Bormann, T., Klug, C.: Dedicated Implementation of Embedded Vision Systems Employing Low-Power Massively Parallel Feature Computation. In: Proc. of the 3rd VIVA-Workshop on Low-Power Information Processing, pp. 1–8 (2002)
8. König, A., Gratz, A.: Advanced methods for the analysis of Semiconductor Manufacturing Process Data. In: Pal, N.R., Jain, L.C. (eds.) Advanced Techniques in Knowledge Discovery and Data Mining, pp. 27–74. Springer, Heidelberg (2005)
9. Marques de Sá, J.P.: Pattern Recognition: Concept, Methods and Applications. Springer, Heidelberg (2001)
10. Fukunaga, K.: Introduction to statistical pattern recognition, 2nd edn. Academic Press Limited, London (1990)
11. http://www.ics.uci.edu/~mlearn/MLRepository.html
12. Kennedy, J., Eberhart, R.C.: Particle Swarm Optimization. In: Proc. of IEEE Int. Conf. on Neural Networks (ICNN), vol. 4, pp. 1942–1948 (1995)
13. Kennedy, J., Eberhart, R.C.: A Discrete Binary Version of The Particle Swarm Algorithm. In: Proc. of Conf. on System, Man, and Cybernetics, pp. 4104–4109 (1997)
14. Agrafiotis, D.K., Cedeñõ, R.C.: Feature Selection for Structure - Activity Correlation Using Binary Particle Swarms. J. Med. Chem. 45, 1098–1107 (2002)
15. Wang, X., Yang, J., Peng, N., Teng, X.: Finding Minimal Rough Set Reducts with Particle Swarm Optimization. Rough Sets, Fuzzy Sets, Data Mining, and Granular Computing, 451–460 (2005)

Visualization and Complexity Reduction of Neural Networks

Tamas Kenesei, Balazs Feil, and Janos Abonyi

Department of Process Engineering, University of Pannonia
Veszprem P.O.Box 158, H-8201 Hungary
abonyij@fmt.uni-pannon.hu
www.fmt.uni-pannon.hu/softcomp

Abstract. The identification of the proper structure of nonlinear neural networks (NNs) is a difficult problem, since these black-box models are not interpretable. The aim of the paper is to propose a new approach that can be used for the analysis and the reduction of these models. It is shown that NNs with sigmoid transfer function can be transformed into fuzzy systems. Hence, with the use of this transformation NNs can be analyzed by human experts based on the extracted linguistic rules. Moreover, based on the similarity of the resulted membership functions the hidden neurons of the NNs can be mapped into a two dimensional space. The resulted map provides an easily interpretable figure about the redundancy of the neurons. Furthermore, the contribution of these neurons can be measured by orthogonal least squares technique that can be used for the ordering of the extracted fuzzy rules based on their importance. A practical example related to the dynamic modeling of a chemical process system is used to prove that synergistic combination of model transformation, visualization and reduction of NNs is an effective technique, that can be used for the structural and parametrical analysis of NNs.

1 Introduction

Nonlinear black-box models have become more and more important not only in research but also in industrial practice. One of the most often used types of black-box models is neural networks (NN). This type of nonlinear models can be used effectively for many purposes including business decision systems [13], engineering and data mining [12]. It consists of simple but strongly connected units called neurons; and generally robust against the failure of single units. Neural networks can be feedforward or recurrent depending on the type of connections. In this paper only feedforward neural networks will be studied.

The main disadvantage of NNs is that they are often too complex and not interpretable. Complexity and interpretability issues are connected with each other: often a relatively simple cross validation method can be used to determine the proper number of hidden neurons but several problems still remain. As Duch showed [5] a simple performance measure is not enough by itself because two NNs with the same performance can have highly different behavior. Other

E. Avineri et al. (Eds.): Applications of Soft Computing, ASC 52, pp. 43–52.
springerlink.com   © Springer-Verlag Berlin Heidelberg 2009

problem is how a priori knowledge can be utilized and integrated into the black-box modeling approach, and how a human expert can validate the identified NNs or more favorably, follow the identification process to interfere in it if it is needed (e.g. to avoid overparameterization or overlook the possible soft or crisp constrains).

To overcome these problems, there are some strategies in the literature:

1. Transformation of NN. The aim of this type of methods is to convert NN into a more interpretable form. Because NN is a black-box, other black-box models should be used that are closer to human thinking. A good approach is to extract rules from NN functions and parameters, and represent them as fuzzy (linguistically sound) if-then rules [2, 13]. For detailed description see Section 2.
2. Model reduction. This approach does not aim to give a 'picture' of NN responses to specific inputs and behavior, but overcome complexity problems with the determination of 'importance' of hidden neurons and weights, remove the insignificant ones, and/or merge the similar ones. It is also important from the viewpoint of overparameterization and to reduce the time and computational demand of the model. Naturally, it can be combined with the above mentioned approaches [3, 8]. For detailed description see Section 3.
3. Visualization of neural network behavior. This approach utilizes the natural pattern recognition capability of human expert. It aims to draw a two dimensional map that is in connection with the behavior of NN in a specific way.[5, 6, 7]. For detailed description see Section 4.

This article introduces how these methods (reduction, visualization and transformation) can be combined into a new effective tool, which can be used for the analysis of NNs. The key idea is that the neural networks can be transformed into a fuzzy rulebase where the rules can be analised.

Section 2 gives a brief introduction and overview about the applied NN transformation method which means the basis for model reduction and visualization. Section 3 contains a combined approach used in this paper to get reduced rule based model from NN. Section 4 overview some NN visualization method, and propose a new technique to measure the similarity of neurons which gives the basis of the visualization approach. In Section 5 some illustrative examples are given, and Section 6 concludes the paper.

2   NN Transformation into Rule Based Model

This subsection gives a brief introduction how NNs work. This description is mainly based on [2], for a detailed discussion see [12]. The multilayer feedforward network has n input variables $(x_1, \ldots, x_n)$, h hidden neurons $(z_1, \ldots, z_h)$, and m outputs $(y_1, \ldots, y_m)$. Let τ_j be the bias for neuron z_j. Let w_{ij} be the weight of the connection from neuron x_i to neuron z_j and β_{jk} the weight of the connection from neuron z_j to y_k. The $\Re^n \to \Re^m$ function, the net calculates is $F(x_1, \ldots, x_n) = (y_1, \ldots, y_m)$ where $y_k = g_A \left(\sum_{j=1}^{h} z_j \beta_{jk} \right)$ with $z_j = f_A \left(\sum_{i=1}^{n} x_i w_{ij} \right)$ where g_A

and f_A are activation functions. In several applications the output activation functions are linear ones, and the usual choice for the hidden activation function is the logistic function: $f_A(x) = 1/(1 + e^{-x})$.

A possible strategy for 'opening' a NN is to convert it into a rule based model. These 'linguistically sound' rules are often fuzzy if-then rules, and are close to human thinking: IF *a set of conditions is satisfied*, THEN *a set of consequences is inferred*. Fuzzy logic provides a tool to process uncertainty, hence fuzzy rules represents knowledge using linguistic labels instead of numeric values, thus, they are more understandable for humans and may be easily interpret [2]. If NNs can be transformed into rules, then it makes possible to overlook and validate the trained NN, and build in a priori knowledge to the network. The crucial question is what the connection is between the several types of neural networks and fuzzy rule based systems.

Under some conditions, the equivalence of normalized radial basis function networks (RBF) and Takagi-Sugeno fuzzy models can be obtained [12]. However, in this paper multilayer perceptron (MLP) type neural networks with logistic hidden activation function are dealt with (in the following the notation NN will be used for MLP type networks). An approach for NNs with tanh activation function is presented in [13] for function approximation purposes, but it should be noted that it is an approximation: the rule based model is not identical to the original trained NN, therefore information transfer in the 'opposite' direction, i.e. from the rule base to the NN can be problematic. An interesting result was given in [2] where the equality of NNs with logistic activation function and a certain type of fuzzy rule based model called fuzzy additive system (FAS) was proven. For that purpose, a new fuzzy logic operator had to be introduced. Because of the equality (which is stronger than equivalence), if a method can be applied on a FAS for a certain purpose (e.g. rule base reduction), then it is also applicable to the NN as well and vice versa. In the following, this equality relation is discussed based on [2]. FAS employs rules in the following form:

$$\mathbf{R}_{jk} : \text{ If } x_1 \text{ is } A_{jk}^1 \text{ and} \dots \text{and } x_n \text{ is } A_{jk}^n \text{ then } y_k \text{ is } p_{jk}(x_1, \dots, x_n) \quad (1)$$

where $p_{jk}(x_1, \dots, x_n)$ is a linear function of the inputs. In FAS's, the inference engine works as follows: for each rule, the fuzzified inputs are matched against the corresponding antecedents in the premises giving the rule's firing strength. It is obtained as the t-norm (usually the minimum operator) of the membership degrees on the rule if-part. The overall value for output y_k is calculated as the weighted sum of relevant rule outputs. Let us suppose multi-intput single-output fuzzy rules, having l_k of them for kth output. Then y_k is computed as $y_k = \sum_{j=1}^{l_k} v_{jk} p_{jk}(x_1, \dots, x_n)$ where v_{jk} is the firing strength of j th rule for k th output.

To decompose the multivariate logistic function to form the rule antecedents in the form of (1) with *univariate* membership functions, a special logic operator has to be used instead of *and* : interactive-or or i-or: $a*b = (a*b)\backslash((1-a)(1-b)+ab)$. Using this $*$ operator, the interpretation of NNs whose hidden neurons have biases as follows. It can be checked that $f_A\left(\sum_{i=1}^{n} x_i w_{ij} + \tau_j\right) = f_A\left(x_1 w_{1j} + \tau_j'\right) * \dots * f_A\left(x_n w_{nj} + \tau_j'\right)$ where $\tau_j' = \tau_j/n$ and the first term corresponds to the fuzzy proposition "$\sum_{i=1}^{n} x_i w_{ij} + \tau_j$ is A". Likewise, $f_A\left(x_i w_{ij} + \tau_j'\right)$ corresponds to proposition "$x_i w_{ij} + \tau_j'$ is A" or in a similar form "$x_i w_{ij}$ is $A - \tau_j'$". Hence, the bias term means a sheer translation. The A_{jk}^i fuzzy sets have to be redefined to account for both the weight w_{ij} and the bias τ_j'. Their membership function is defined by: $\mu_{A_{jk}^i}(x) = \mu_A\left[(x + \tau_j') * w_{ij}\right]$. Based on that, the fuzzy rules extracted from the trained NN are:

$$\mathbf{R}_{jk} : \mathbf{If} \ x_1 \ is \ A_{jk}^1 \ * \ldots and \ * \ x_n \ is \ A_{jk}^n \ \mathbf{then} \ y_k \ = \ \beta_{jk} \qquad (2)$$

3   Model Complexity Reduction

In this section we focus on the combination of existing model reduction techniques with the previously presented rule based model extraction method. An interesting solution to NN reduction is the following: the complexity of the model is penalized, and it is built-in to the training procedure. The method proposed in [3] uses a cost function that consists of two terms: one for the NN accuracy (like mean square error) and one related to the NN complexity (numbers and magnitude of parameters). However, determination of their weights or relative importance is problematic. A weighting factor is introduced and several NNs should be trained with different weighting parameters. To compare the trained NNs and choose the best one, [3] applied the predicted square error measure.

Another possible and often applied solution is to prune the identified model trained with classical cost function. In the following, model reduction techniques of this type will be considered. In general it can be stated that linear model reduction methods are preferred to nonlinear ones because they are exhaustively studied and effectively applied for several types of problems. For that purpose the model should be linear in parameters. A possible method family is orthogonal techniques. These methods can roughly be divided into two groups: the rank revealing ones like SVD-QR algorithm and those that evaluate the individual contribution of the rule or local models, like the orthogonal least-squares approach (OLS). This later technique requires more computations, but for system identification purposes it is preferable as it gives a better approximation result. In the remaining part of this paper OLS is applied for rule ranking and model reduction purposes. OLS works as follows (for a throughout discussion see [12]). Consider a general linear in parameters model:

$$\mathbf{y} = \mathbf{Z}\theta + \mathbf{e} \qquad (3)$$

where $\mathbf{y} = [y_1, \ldots, y_N]^T$ is the measured output, $\mathbf{Z} = [\mathbf{z}_1, \ldots, \mathbf{z}_n]^T$ is the regressor matrix ($\mathbf{z_i} = [z_{i1}, \ldots, z_{iN}]^T$, $i = 1, \ldots, h$ are the regressors) $\theta = [\theta_1, \ldots, \theta_h]$ is the parameter vector and $\mathbf{e} = [e_1, \ldots, e_N]^T$ is the prediction error. OLS transforms the columns of the regressor matrix $\mathbf{Z}$ into a set of orthogonal basis vectors in order to inspect the individual contribution of each regressor. If they were not orthogonal, they could not been inspected individually. An orthogonalization method should be used to perform the orthogonal decomposition

$\mathbf{Z} = \mathbf{VR}$ (often the simple Gram-Schmidt method is used), where $\mathbf{V}$ is an orthogonal matrix such that $\mathbf{V}^T\mathbf{V} = \mathbf{I}$ and $\mathbf{R}$. Substituting $\mathbf{Z} = \mathbf{VR}$ into (3), we get $\mathbf{y} = \mathbf{VR}\theta + \mathbf{e} = \mathbf{Vg} + \mathbf{e}$. where $\mathbf{g} = \mathbf{R}\theta$. Since the columns $\mathbf{v_i}$ of $\mathbf{V}$ are orthogonal, the sum of squares of y_k can be written as

$$\mathbf{y}^T\mathbf{y} = \sum_{i=1}^{h} g_i^2 \mathbf{v}_i^T \mathbf{v}_i + \mathbf{e}^T\mathbf{e} \tag{4}$$

The part of the output variance $\mathbf{y}^T\mathbf{y}/N$ explained by regressors is $\sum_{i=1}^{h} g_i^2 \mathbf{v}_i^T \mathbf{v}_i/N$ and an error reduction ratio due to an individual regressor i can be defined as

$$err_i = \frac{g_i^2 \mathbf{v}_i^T \mathbf{v}_i}{\mathbf{y}^T\mathbf{y}}, \quad i = 1, \ldots, h. \tag{5}$$

This ratio offers a simple means of ordering the regressors. As [12] shows, "there are only two restrictions to the application of this subset selection technique. First, the model has to be linear in parameters. Second, the set of regressors from which the significant ones will be chosen must be precomputed." This later one is an important restriction because it means that all regressors are fixed during this procedure. By normalized RBF networks and Takagi-Sugeno fuzzy models this requirement is not met, therefore the original version of OLS cannot be applied. It is because the normalization denominator changes as the number of selected rules changes, thus the fuzzy basis functions (here: regressors) change. To overcome this problem the value of the denominator can be fixed, but in this case interpretability issues are discarded completely. However, OLS can be very useful for various purposes; modified versions of OLS can also be applied to determine the centers of radial basis functions, or to generate Takagi-Sugeno-Kang fuzzy models.

In case of MLP networks and FAS systems, this problem does not occur because of the special output computing mechanism. Thus classical OLS can be applied on FAS systems to rank the rules since the parameters of the trained NN are fixed. However, OLS is formulated as a MISO technique. If the NN has more than one output, then the outputs can be evaluated individually one by one. In this case (using the notation of OLS (3)-(5)), $\mathbf{y}$ is the kth network output, the regressors $\mathbf{z}_i$ are the outputs of the hidden neurons, and the parameters θ_j corresponds to the weights from the jth hidden neuron to the kth output neuron β_{jk}. This approach was directly applied on NNs in [9], and it was shown that analog method can be applied to the subset selection of the original network inputs. In this case in (3)-(5), $\mathbf{y}$ is the output of the kth hidden neuron, the regressors $\mathbf{z}_i$ are the inputs of the network, and the parameters θ_j corresponds to the weights from the jth input neuron to the kth hidden neuron w_{jk}. Other NN pruning can also be considered, e.g. optimal brain damage [4] or optimal brain surgeon [8], and it should be emphasized that these methods can directly be applied on FAS systems as well. The application examples in Section 5 show that it can be very effective if a model reduction technique (in this paper OLS for ranking the rules) and rule base extraction from NN are applied together, and validate the identified models by human experts.

Note that ordering the neurons by OLS estimated error reduction ratios reveals the unnecessary neurons (the importance of the extracted rules) in the hidden layer, because neurons with low error reduction ratio are insignificant for the appropriate model. As the equality of FAS and NN was proven in [2] and was discussed also in Section 2, the OLS ranking means a reduction based on the *consequent of the fuzzy rule.*

It should be noted that the applied *i*-or operator in the extracted fuzzy rules does not belong to the commonly applied fuzzy *t*-norms or *t*-conorms. However, it would be interesting to test the extracted fuzzy rules with common fuzzy logic operators, and maybe recompute the output weights (which can easily be done because the model is linear with respect to these parameters). Our presumption is that the crisper the activation functions are (f_A), the less the difference is between the modeling performances of the original and the modified FAS's that uses classical fuzzy logic operators. For that purpose, numerous tests have to be completed in the future. If this guess proves true, then the cost function for training the NN can be modified to get 'crisper' activation functions.

4   NN Visualization Methods

In this section, a new technique for the visualization of neural networks is proposed. First, methods are discussed that can directly be applied on NNs. Second, a new approach is presented to detect the redundant neurons based on their similarity. This method exploits the equality of NNs and FAS's because it is based on the similarity of fuzzy membership functions.

The output of the hidden neurons z_j can be seen as a h dimensional vector that represents the range the neurons work in. If a 'hidden variable' z_j is close to zero or one, the neuron is saturated. If a hidden neuron gives values near zero or one for almost all inputs, hence it does not fire or fires all the time, it is useless for the problem. The distribution of these h dimensional data can represent the NN behavior for a human expert. Unfortunately, in several cases there is a need for more than two or three hidden neurons. In these cases a projection or dimensionality reduction technique has to be used. Principal Component Analysis (PCA) is a linear technique; therefore the information loss may be more than the admissible level. Other (topology or distance preserving) projection techniques like Multidimensional Scaling, Sammon method, Isomap or Locally Linear Embedding can be used for that purpose. For more details see [1] and the references within.

However, there are some special visualization methods for NNs. Duch [5] proposed an approach for visualization of NNs applied on classification problems. His method can be applied for problems with K classes if the output is coded as a K length vector: $(1, 0, \ldots, 0)$ means the first class, $(0, 1, \ldots, 0)$ the second and so on. In this case case the classes are represented by the corners of the K dimensional unit hypercube. The approach proposed by Duch maps the NN *output* into two dimensions, basically 'flattens' the hypercube into two dimensions. This approach was thought over and applied on the *output of the hidden neurons*

z_j in [6, 7]. This method was straightforward from the former one because the hidden variables (the activation functions) take values from $[0, 1]$, therefore the h dimensional vectors are located within the unit hypercube. This method can be used not only for classification but also for function approximation purposes as well. Based on this latter approach a picture of the behavior of the hidden units, their firing strength and activation or saturation level can be obtained. The main drawback is that the number of classes/hidden neurons is limited. To keep the figures simple and interpretable, only $3 \ldots 6$ variables can be used.

In the following, a different method is proposed to visualize the NN. The presented approach utilizes the *antecedent part* of the extracted fuzzy rules (since OLS based model reduction uses the consequent parts, see Section 3). To reduce the FAS rule base by analyzing the antecedent part of the rules is possible with measuring the similarity of the membership functions, and removing the too similar neurons. Utilizing the equality of FAS and NN, the following classical interclass separability measure (originally for fuzzy systems) could be used to compare the univariate functions decomposed from hidden neurons:

$$S_{j,l,k}^i = \frac{\int \min\left(A_{j,k}^i(x_i), A_{l,k}^i(x_i)\right) dx_i}{\int \max\left(A_{j,k}^i(x_i), A_{l,k}^i(x_i)\right) dx_i} \tag{6}$$

where $i = 1, \ldots, n$, $j, l = 1, \ldots, h..$ Eq. (6) can be used to measure the similarity of two *clauses* in the rule base, in other words the similarity of two hidden neurons for the same input variables. To compare the hidden neurons themselves with multivariate activation functions, the following measure seems to be straightforward:

$$S_{j,l,k} = \prod_i S_{j,l,k}^i, \ i = 1, \ldots, n, \ j, l = 1, \ldots, h. \tag{7}$$

With this measure, pairwise similarities of hidden neurons in the range of $[0, 1]$ can be obtained. To get pairwise distances if needed, the simple form of $1 - S_{j,l,k}$ can be used. Based on these distances which can be called relative data, the neurons themselves can be mapped onto two dimensions. In this paper, the classical multidimensional scaling will be used. This well-known technique is not discussed here because it would exceed the size and scope of this paper. The mapped two dimensional points refer how similar the neurons behave. As can be seen, the above mentioned approaches [6, 7] visualize the output of the hidden neurons, and draw conclusions from the location of these data. The proposed approach focuses to the behavior of hidden neurons as well, but utilizes the shape of the identified multidimensional activation functions. The previous approach can be used to determine how well the NN was trained, since the proposed one show which neurons are similar and redundant within the trained network. In this formulation, this method can rather be used for complexity reduction purposes, and not to qualify the training procedure.

5   Application Examples and Discussion

For applying the intruced visualization and reduction techniques we used a
dataset of a pH process, where the concentration of hydrogen ions in a continuous stirred tank reactor is modeled (CSTR). This well-known modeling problem
presents difficulties due to the nonlinearity of the process dynamics. This process
can be correctly modeled as a first-order input-output (NARX) system, where
the actual output (the pH) $y(k+1)$, depends on pH of the reactor $y(k)$ and the
NaOH feed $u(k)$ at the kth sample time (sample time is $t_s = 0.2min$):

$$y(k+1) = f(y(k), u(k)) \qquad (8)$$

Parameters of the neural network were identified by the back-propagation
algorithm based on a uniformly distributed training data where F_{NaOH} is in the
range of 515-525 l/min. Our experiences show that 7 neurons are sufficient in
the hidden layer of the NN.

Applying the proposed visualization and transformation techniques, Figure 1
shows the decomposed univariate membership functions and the into two dimension mapped distance matrix according to the NN model parameters. For better
interpretability, the histogram of the corresponding model inputs are illustrated
on the last two subplots. On the right the pairwise distances of the neurons (see
(6) in the previous section) were mapped into two dimensions with MDS and
two dimensional points refer how similar the neurons behave.

The neurons are listed according to the OLS ranking on the left of Figure 1,
starting with the rules decomposed from the most important neuron in the

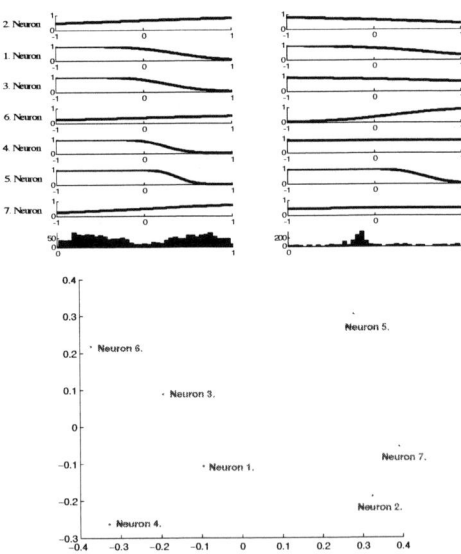

Fig. 1. Decomposed univariate membership functions(left), Distances between neurons
mapped into two dimensions with mds (right)

Table 1. Training errors for different model structures and model reductions. (Number of neurons in the hidden layer of the network/Number of reduced neurons from the given model structure).

-	6	7	8	9	10	11	12	13	14	15
1	$6.8 \cdot 10^{-4}$	$4.4 \cdot 10^{-5}$	$2.7 \cdot 10^{-5}$	$3.4 \cdot 10^{-5}$	$4.5 \cdot 10^{-5}$	$2.9 \cdot 10^{-5}$	$3.3 \cdot 10^{-5}$	$6.6 \cdot 10^{-4}$	$3.6 \cdot 10^{-5}$	$6.6 \cdot 10^{-4}$
2	$4.8 \cdot 10^{-5}$	$6.7 \cdot 10^{-4}$	$3.9 \cdot 10^{-5}$	$3.9 \cdot 10^{-5}$	$6.7 \cdot 10^{-4}$	$3.4 \cdot 10^{-5}$	$3.3 \cdot 10^{-5}$	$3.6 \cdot 10^{-5}$	$5.9 \cdot 10^{-5}$	$1.1 \cdot 10^{-4}$
3	$6.5 \cdot 10^{-5}$	$3.8 \cdot 10^{-5}$	$1.7 \cdot 10^{-5}$	$1.4 \cdot 10^{-4}$	$3.1 \cdot 10^{-5}$	$2.7 \cdot 10^{-5}$	$8.9 \cdot 10^{-5}$	$3.3 \cdot 10^{-5}$	$3.5 \cdot 10^{-5}$	$3.8 \cdot 10^{-5}$
4	$6.4 \cdot 10^{-5}$	$3.3 \cdot 10^{-5}$	$3.6 \cdot 10^{-5}$	$5.0 \cdot 10^{-5}$	$3.2 \cdot 10^{-5}$	$6.6 \cdot 10^{-4}$	$5.8 \cdot 10^{-5}$	$3.9 \cdot 10^{-5}$	$3.3 \cdot 10^{-5}$	$3.8 \cdot 10^{-5}$
5	$7.7 \cdot 10^{-5}$	$2.3 \cdot 10^{-5}$	$5.3 \cdot 10^{-5}$	$3.4 \cdot 10^{-5}$	$3.5 \cdot 10^{-5}$	$3.8 \cdot 10^{-5}$	$3.4 \cdot 10^{-5}$	$3.8 \cdot 10^{-5}$	$3.5 \cdot 10^{-5}$	$4.2 \cdot 10^{-5}$
6	$- -$	$2.9 \cdot 10^{-5}$	$4.4 \cdot 10^{-5}$	$5.6 \cdot 10^{-5}$	$5.2 \cdot 10^{-5}$	$6.3 \cdot 10^{-4}$	$6.5 \cdot 10^{-4}$	$3.1 \cdot 10^{-5}$	$8.4 \cdot 10^{-5}$	$7.2 \cdot 10^{-5}$
7	$- -$	$- -$	$6.4 \cdot 10^{-4}$	$4.4 \cdot 10^{-5}$	$3.5 \cdot 10^{-5}$	$6.2 \cdot 10^{-5}$	$4.1 \cdot 10^{-5}$	$5.2 \cdot 10^{-5}$	$3.4 \cdot 10^{-5}$	$4.4 \cdot 10^{-5}$
8	$- -$	$- -$	$- -$	$1.7 \cdot 10^{-5}$	$6.1 \cdot 10^{-5}$	$7.9 \cdot 10^{-5}$	$3.7 \cdot 10^{-5}$	$3.2 \cdot 10^{-5}$	$3.6 \cdot 10^{-5}$	$7.1 \cdot 10^{-5}$
9	$- -$	$- -$	$- -$	$- -$	$4.8 \cdot 10^{-5}$	$4.4 \cdot 10^{-5}$	$1.7 \cdot 10^{-5}$	$4.9 \cdot 10^{-5}$	$3.7 \cdot 10^{-5}$	$4.7 \cdot 10^{-5}$
10	$- -$	$- -$	$- -$	$- -$	$- -$	$3.2 \cdot 10^{-5}$	$3.2 \cdot 10^{-5}$	$8.7 \cdot 10^{-5}$	$9.0 \cdot 10^{-5}$	$1.2 \cdot 10^{-4}$
11	$- -$	$- -$	$- -$	$- -$	$- -$	$- -$	$3.1 \cdot 10^{-5}$	$7.0 \cdot 10^{-5}$	$5.0 \cdot 10^{-5}$	$1.5 \cdot 10^{-4}$
12	$- -$	$- -$	$- -$	$- -$	$- -$	$- -$	$- -$	$3.6 \cdot 10^{-5}$	$3.3 \cdot 10^{-5}$	$4.5 \cdot 10^{-5}$
13	$- -$	$- -$	$- -$	$- -$	$- -$	$- -$	$- -$	$- -$	$3.1 \cdot 10^{-5}$	$1.2 \cdot 10^{-4}$
14	$- -$	$- -$	$- -$	$- -$	$- -$	$- -$	$- -$	$- -$	$- -$	$2.6 \cdot 10^{-5}$

hidden layer of the network. The consequence of synthesizing the results is that it is possible to remove one neuron out of 7 (in FAS the corresponding rules) from the model without a significant increase in modeling performance, because of the low error reduction rate of the last, 7th neuron. This achievement harmonizes with the issues of the mapped distances, where the 2nd and the 7th neuron are closer to each other, but OLS based ranking indicates the 2nd one as more important.

Model reduction and visualization techniques like OLS makes it possible to overcome the problem of overfitting and the performance of the reduced model is almost the same as the original one. A rigorous test of NARX models is free run simulation because the errors can be cumulated. The result indicates the goodness of the reduced model even by free run simulation ($3.5 \cdot 10^{-3}$ for neural network with 7 neurons, $3.823 \cdot 10^{-3}$ using i-or for FAS with 7 rules, $3.824 \cdot 10^{-3}$ after removing 1 neuron from the hidden layer containing 7 neurons).

Table 1 shows the training errors for neural networks with different number of neurons in the hidden layer($6 - 15$). These models were reduced with $1 - 14$ neurons. The obtained results points on that it is worth considering to select the appropriate model structure because better modeling performance can be achieved with reducing an overfitted model. The reason of this phenomena is that the gradient based training algorithms may stop in different local minimums.

The used similarity measure can be applied for further reduction of the rule base. It can be done in an automatic way if a threshold value is defined previously. If the measured similarity is greater than the threshold, the corresponding two neurons in the original neural network can be considered as identical; therefore further reduction of the FAS rule base is possible. This technique can be used even in the learning process of the neural network.

6   Summary

Neural networks are often too complex and not interpretable, therefore it is very difficult to utilize these networks correctly. This article proposed a new complex approach for visualization and reduction of the neural networks, and discussed that neural network with sigmoid transfer function is identical to fuzzy additive systems.

A possible future research area is to develop a new learning procedure for neural networks using prior knowledge based if-then rules, which combines the user's experience and/or constraints with the learning capability of NN.

Acknowledgments

The authors would like to acknowledge the support of the Cooperative Research Centre (VIKKK, project III/1) and Hungarian Research Found (OTKA T049534). Janos Abonyi is grateful for the support of the Bolyai Research Fellowship of the Hungarian Academy of Sciences and the veges Fellowship.

References

1. Abonyi, J., Feil, B.: Aggregation and Visualization of Fuzzy Clusters based on Fuzzy Similarity Measures. In: Advances in Fuzzy Clustering and its Applications, pp. 95–123. John Wiley & Sons, Chichester (2007)
2. Benitez, J.M., Castro, J.L., Requena, I.: Are artifical neural networks black boxes? IEEE Transactions on Neural Networks 8(5), 1156–1164 (1997)
3. Bath, N.V., McAvoy, T.J.: Determining model structure for neural models by network stripping. Computers chem. Engng. 16(4), 271–281 (1992)
4. Cun, Y.L., Denker, J., Solla, S.: Optimal brain damage. Advances in neural information processing systems 2, 598–605 (1990)
5. Duch, W.: Coloring black boxes: visualization of neural network decisions. In: Int. Joint Conf. on Neural Networks, Portland, Oregon, vol. 1, pp. 1735–1740. IEEE Press, Los Alamitos (2003)
6. Duch, W.: Visualization of Hidden Node Activity in Neural Networks: I. In: Rutkowski, L., Siekmann, J.H., Tadeusiewicz, R., Zadeh, L.A. (eds.) ICAISC 2004. LNCS, vol. 3070, pp. 38–43. Springer, Heidelberg (2004)
7. Duch, W.: Visualization of hidden node activity in neural networks: II. In: Rutkowski, L., Siekmann, J.H., Tadeusiewicz, R., Zadeh, L.A. (eds.) ICAISC 2004. LNCS, vol. 3070, pp. 44–49. Springer, Heidelberg (2004)
8. Hassibi, B., Stork, D., Wolff, G.: Optimal brain surgeon and general network pruning. Technical Report 9235, RICOH California Research Center, Menlo Park, CA (1992)
9. Henrique, H.M., Lima, E.L., Seborg, D.E.: Model structure determination in neural network models. Chemical Engineering Science 55, 5457–5469 (2000)
10. Linkens, D.A., Chen, M.-Y.: Input selection and partition validation for fuzzy modelling using neural network. Fuzzy Sets and Systems 107, 299–308 (1999)
11. Mitra, S., Pal, S.K.: Fuzzy multi-layer perceptron, inferencing and rule generation. IEEE Transactions on Neural Networks 6(1), 51–63 (1995)
12. Nelles, O.: Nonlinear system identification. Springer, Heidelberg (2001)
13. Setiono, R., Leow, W.K., Thong, J.Y.L.: Opening the neural network black box: an algorithm for extracting rules from function approximating artificial neural networks. In: Proceedings of the 21st International Conference on Information systems, Queensland, Australia, pp. 176–186 (2000)

## Part II

Modelling, Control and Optimization

Artificial Immune Network Combined with Normative Knowledge for Power Economic Dispatch of Thermal Units

Leandro dos Santos Coelho[1] and Viviana Cocco Mariani[2]

[1] Industrial and Systems Engineering Graduate Program, LAS/PPGEPS, Pontifical Catholic University of Paraná, PUCPR, Imaculada Conceição, 1155, Zip code 80215-901, Curitiba, Paraná, Brazil
[2] Mechanical Engineering Graduate Program, PPGEM, Pontifical Catholic University of Paraná, PUCPR, Imaculada Conceição, 1155, Zip code 80215-901, Curitiba, Paraná, Brazil

Summary. Recently, many research activities have been devoted to Artificial Immune Systems (AISs). AISs use ideas gleaned from immunology to develop intelligent systems capable of learning and adapting. AISs are optimization methods that can be applied to the solution of many different types of optimization problems in power systems. In particular, a new meta-heuristic optimization approach using artificial immune networks called opt-aiNET combined with normative knowledge, a cultural algorithm feature, is presented in this paper. The proposed opt-aiNET methodology and its variants are validated for a economic load dispatch problem consisting of 13 thermal units with incremental fuel cost function takes into account the valve-point loadings effects. The proposed opt- aiNET approach provides quality solutions in terms of efficiency compared with other existing techniques in literature for load dispatch problem with valve-point effect.

1 Introduction

The economic dispatch problem (EDP) is to determine the optimal combination of power outputs of all generating units to minimize the total fuel cost while satisfying the load demand and operational constraints. It plays an important role in operation planning and control of modern power systems. Under a new deregulated electricity industry, power utilities try to achieve high operating efficiency to produce cheap electricity. Therefore, precise generation costs analysis [1] and several constraints, such as transmission capacity and system security are very important issues in the economic dispatch problems. Due to the nature of large scale, nonlinear generation cost and multiple constraints, EDP problem combines a highly nonlinear, nonconvex and computationally difficult environment with a need for optimality [2].

Over the past few years, a number of approaches have been developed for solving the EDP using the dynamic programming, linear programming, homogenous linear programming, and nonlinear programming techniques [3],[4]. In these numerical methods for solving the EDP, an essential assumption is that the

E. Avineri et al. (Eds.): Applications of Soft Computing, ASC 52, pp. 55–64.
springerlink.com   © Springer-Verlag Berlin Heidelberg 2009

incremental cost curves of the units are piecewise-linear monotonically increasing functions. Unfortunately, the input-output characteristics of modern power generating units are inherently highly nonlinear because of valve-point loadings, multi-fuel effects, and others. Furthermore, they may lead to multiple local minimum points of the cost function. Classical dispatch algorithms require that these characteristics be approximated, even though such approximations are not desirable as they may lead to suboptimal solutions and hence huge revenue losses over time [5]. These traditional methods based on gradient information suffer from myopia for nonlinear, discontinuous search spaces, leading them to a less-that desirable performance, often use approximations to limit complexity. When search space is particularly irregular (due to inclusion of valve-point effects), algorithms need to be highly robust to escape form premature convergence [6].

In this context, as an alternative to the conventional mathematical approaches, the bio-inspired intelligent systems, such as evolutionary algorithms and artificial neural networks [7]–[12] have been given much attention by many researchers due to their ability to seek for the near global optimal solution without any restrictions in the shape of the cost curves. One of these modern heuristic optimization paradigms is the artificial immune system (AIS).

AIS is a very intricate biological system which accounts for resistance of a living body against harmful foreign entities. Artificial immune system (AIS) imitates the immunological ideas to develop some techniques used in various areas of research [13],[14].

A meta-heuristic optimization approach employing AIS called opt-aiNET algorithm to solve the EDP is proposed in this paper. The aiNET algorithm is a discrete immune network algorithm based on the artificial immune systems paradigm that was developed for data compression and clustering [15], and was also extended slightly and applied to optimization to create the algorithm opt-aiNET [16]. Opt-aiNET, proposed in [15], evolves a population, which consists of a network of antibodies (considered as candidate solutions to the function being optimized). These undergo a process of evaluation against the objective function, clonal expansion, mutation, selection and interaction between themselves.

An EDP with 13 unit test system using nonsmooth fuel cost function [17] is employed in this paper for demonstrate the performance of the proposed opt-aiNET method and a new approach of opt-aiNET combined with normative knowledge, a feature of cultural algorithms. The cultural algorithm was introduced by Reynolds as a vehicle for modeling social evolution and learning in agent based societies [18],[19]. A cultural algorithm consists of an evolutionary population of agents whose experiences are integrated into a belief space consisting of various forms of symbolic knowledge. The cultural algorithm framework easily lends itself to supporting various types of learning activities in the belief space, including ensemble learning [20].

The results obtained with the opt-aiNET approaches were analyzed and compared with those obtained in recent literature.

2 Formulation of Economic Dispatch Problem

The objective of the economic dispatch problem is to minimize the total fuel cost at thermal power plants subjected to the operating constraints of power system. Therefore, it can be formulated mathematically with an objective function and two constraints. The equality and inequality constraints are represented by equations (1) and (2) given by

$$\sum_{i=1}^{n} P_i - P_L - P_D = 0 \tag{1}$$

$$P_i^{min} \leq P_i \leq P_i^{max} \tag{2}$$

In the power balance criterion, an equality constraint should be satisfied, the equation (1). The generated power should be the same as total load demand plus the total line losses. The generation power of each generator should be laid between maximum and minimum limits represented by equation (2), where P_i is the power of generator i (in MW); n is generators number present in the system; P_D is the total system demand (in MW); P_L is the total lines losses (in MW) and P_i^{min} and P_i^{max} are, respectively, the outlet of the operation minimum and maximum of the generator unit i (in MW). The total fuel cost function is formulated as follows

$$\min f = \sum_{i=1}^{n} F_i(P_i) \tag{3}$$

where F_i is the total fuel cost for the generator unity i (in \$/h), that is defined by equation,

$$F_i(P_i) = a_i P_i^2 + b_i P_i + c_i \tag{4}$$

where a_i, b_i and c_i are cost coefficients of generator i.

A cost function is obtained based on the ripple curve for more accurate modeling. This curve contains higher order non-linearity and discontinuity due to valve point effect, and should be refined by sine function. Therefore, the equation (4) can be modified [3] as

$$\tilde{F}_i(P_i) = F(P_i) + \left| e_i \sin \left(f_i \left(P_i^{min} - P_i \right) \right) \right| \tag{5}$$

or

$$\tilde{F}_i(P_i) = a_i P_i^2 + b_i P_i + c_i + \left| e_i \sin \left(f_i \left(P_i^{min} - P_i \right) \right) \right| \tag{6}$$

where e_i and f_i are constants from the valve point effect of generators. Consequently, the total fuel cost that must be minimized, accordant represented in the equation (3), is modified for

$$\min f = \sum_{i=1}^{n} \tilde{F}_i(P_i) \tag{7}$$

where $\tilde{F}_i$ is the cost function of generator i (in \$/h) defined by equation (6). In the case study presented in this paper is disregarded the transmission losses, P_L, thus $P_L = 0$.

3   Optimization Methodology Based on opt-aiNET for EDP

Opt-aiNET is capable of performing local and global search, as well as to adjust dynamically the size of the population [24]. Opt-aiNET creates a memory set of antibodies (points in the search space) which represent (over time) the best candidate solutions to the objective function minimization problem. Opt-aiNET is capable of either unimodal or multimodal optimization and is characterized by five main features: (i) its population size is dynamically adjustable; (ii) it naturally combines exploitation and exploration of the search space; (iii) it determines the locations of multiple optima; (iv) it has the capability of maintaining many optima solutions; and (v) it has defined stopping criteria. The steps of opt-aiNET can be summarized as follows:

Initialization of the parameter setup
The user must choose the key parameters that control the opt-aiNET, i.e., population size (M), suppression threshold (σ_s), number of clones generated for each cell (N_c), percentage of random new cells each iteration (d), scale of affinity proportion selection (β), and maximum number of iterations allowed (stopping criterion), N_{gen}.

Initialization of cell populations
Set iteration $t = 1$. Initialize a population $c_{ij}(t)$ of $i = 1, \ldots, M$ cells (where $j = 1, \ldots, n$-dimensional solution vectors) with random values generated according to a uniform probability distribution in the n dimensional problem space. Initialize the entire solution vector population in the given upper and lower limits of the search space.

Evaluation of each network cell
Evaluate the fitness value of each cell.

Generation of clones
Generate a number N_c of clones for each network cell. The clones are offspring cells that are identical copies of their parent cell.

Mutation operation
Mutation is an operation that changes each clone proportionally to the fitness of the parent cells, but keeps the parent cell. Clones of each cell are mutated according to the affinity (Euclidean distance between two cells applied in the phenotypic space) of the parent cell. The affinity proportional mutation is performed according to equations (8) and (9), given by:

$$c_{ij}(t+1) = c_{ij}(t) + \alpha N(0,1) \tag{8}$$

$$\alpha = \rho^{-1}e^{-f*} \tag{9}$$

where $c_{ij}(t+1)$ is a mutated cell, $N(0,1)$ is a Gaussian random variable of zero mean and unitary variance, ρ is a parameter that controls the decay of the inverse exponential function, and $f*$ is the objective function of the parent cell normalized in the range $[0,1]$.

Evaluation the objective function of all network cells
Evaluate the objective function value of all network cells of the population including new clones and mutated clones.

Selection of fittest clones
For each clone select the most fit and remove the others.

Determination of affinity of all network cells
Determine the affinity of network cells and perform network suppression.

Generate randomly d network cells
Introduce a percentage d of randomly generated cells. Set the generation number for $t = t+1$. Proceed to step of *Evaluation of each network cell* until a stopping criterion is met, usually a maximum number of iterations, t_{max}. The stopping criterion depends on the type of problem.

4   Optimization Based on opt-aiNET and Normative Knowledge for EDP

Cultural evolution is driven by the common beliefs that emerge upon the collective experience of selected potential solutions. This belief space knowledge can be viewed as falling into five basic categories: normative, situational, topographical, domain, and historical [20].

Cultural algorithm methods present three components [21],[22]. First, there is a population component (or population space) that contains the population to be evolved and the mechanisms for its evaluation, reproduction and modification in this work, the cloness population in opt-aiNET. Second, there is a belief space that represents the bias that has been acquired by the population during its problem-solving process. The third component is the communication protocol which is used to determine the interaction between the population and the beliefs [23].

Initially, cultural algorithms were applied with population spaces based on evolutionary programming [21] approaches to real parameter optimization. In this work, in opt-aiNET design, new concepts of optimization are presented based on normative knowledge of cultural algorithms. The normative knowledge is viewed as a knowledge source by interacting with other clones in the belief space, produces an ensemble that drives the evolution in the population space.

Cultural algorithms model two levels of evolution: the social population level and the belief space level. In addition to a population space, a cultural algorithm

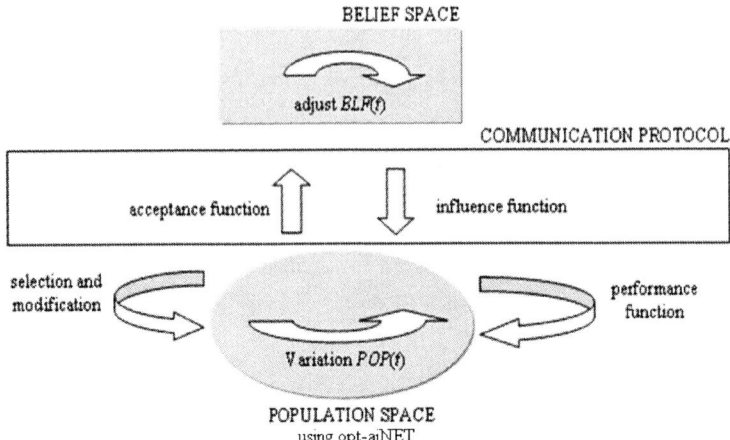

Fig. 1. A framework of cultural algorithm using opt-aiNET

has a belief space in which the beliefs (problem-solving knowledge) acquired from the populations evolution can be stored and integrated. An acceptance function is used to generate beliefs by gleaning the experience of individuals from the population space. In turn, this problem-solving knowledge can bias the evolution of the population component through the influence function. The belief space itself also evolves through the adjust function. A framework of cultural algorithm

```
Notation
% l_i e u_i are the lower and upper of belief space for
  j-th dimension
% i is the index of i-th clone in population of M clones
% c_ij is the i-th clone of j-th dimension

Pseudo-code
Calculate η_accepted clones of population of M clones

Choice the η_accepted of k = {1 v 2 v ...v M}

Calculate l_j and u_j of η_accepted clones

If c_ij(t) < l_j then c_ij(t+1) = l_j + α·|N(0,1)| ;

    Elseif c_ij(t) > u_j then c_ij(t+1) = u_j - α·|N(0,1)| ;

    Else c_ij(t+1) = c_ij(t) + α·N(0,1) ;     % equation (8)

End if
```

Fig. 2. Pseudo-code of opt-aiNET approach based on normative knowledge

is presented in Figure 1. The proposed approach use opt-aiNET as the population space.

Figure 2 shows a pseudo-code of the opt-aiNET with normative knowledge (N-opt-aiNET). The number of particles accepted, $\eta_{accepted}$, for update of the belief space is a design parameter. In this work, $\eta_{accepted}$ is 80% of number of clones, N_c.

5 Description and Simulation Results for the Case Study of 13 Thermal Units

This case study consisted of 13 thermal units of generation with the effects of valve-point loading, as given in Table 1. The data shown in Table 1 are also available in [17] and [25]. In this case, the load demand expected to be determined was $P_D = 1800$ MW.

Each optimization method based on opt-aiNET was implemented in Matlab (MathWorks). All the programs were run on a 3.2 GHz Pentium IV processor with 2 GB of random access memory. In each case study, 30 independent runs were made for each of the optimization methods involving 30 different initial trial solutions for each opt-aiNET approach.

The simulation results obtained are given in Table 2, which shows that the N-opt-aiNET succeeded in finding the best solution for the tested methods. The best results obtained for solution vector P_i, $i = 1, \ldots, 13$ using N-opt-aiNET was 17977.0905 \$/h, as shown in Table 3.

Table 4 compares the results obtained in this paper with those of other studies reported in the literature. Note that in studied case, the result reported here using N-opt-aiNET is comparatively lower than recent studies presented in the literature, except to the simulation results using hybrid particle swarm with SQP proposed in [26].

Table 1. Data for the 13 thermal units

Thermal unit	P_i^{min}	P_i^{max}	a	b	c	e	f
1	0	680	0.00028	8.10	550	300	0.035
2	0	360	0.00056	8.10	309	200	0.042
3	0	360	0.00056	8.10	307	150	0.042
4	60	180	0.00324	7.74	240	150	0.063
5	60	180	0.00324	7.74	240	150	0.063
6	60	180	0.00324	7.74	240	150	0.063
7	60	180	0.00324	7.74	240	150	0.063
8	60	180	0.00324	7.74	240	150	0.063
9	60	180	0.00324	7.74	240	150	0.063
10	40	120	0.00284	8.60	126	100	0.084
11	40	120	0.00284	8.60	126	100	0.084
12	55	120	0.00284	8.60	126	100	0.084
13	55	120	0.00284	8.60	126	100	0.084

Table 2. Convergence results (30 runs) of a case study of 13 generating units with valve point and $P_D = 1800$ MW

Optimization Method	Minimum Cost ($/h)	Mean Cost ($/h)	Maximum Cost ($/h)	Std. Deviation of Cost ($/h)
opt-aiNET	18110.1024	18258.6736	18403.5644	71.4153
N-opt-aiNET	17977.0905	18193.4017	18373.0115	88.3068

Table 3. Best result (30 runs) obtained using N-opt-aiNET approach

Power	Generation (MW)	Power	Generation (MW)
P_1	628.3197	P_8	109.8659
P_2	74.4172	P_9	109.3446
P_3	298.6458	P_{10}	40.0000
P_4	60.0000	P_{11}	40.0000
P_5	109.6849	P_{12}	55.0000
P_6	109.8553	P_{13}	55.0000
P_7	109.8665	$\sum_{i=1}^{n} P_i$	1800.0000

Table 4. Comparison of results for fuel costs presented in the literature

Optimization Technique	Case Study with 13 Thermal Units
Evolutionary programming [17]	17994.07
Particle swarm optimization [26]	18030.72
Hybrid evolutionary programming with SQP [26]	17991.03
Hybrid particle swarm with SQP [26]	17969.93
Proposed N-opt-aiNET approach	17977.0905

6   Conclusion and Future Research

This paper presents opt-aiNET approaches for solving an EDP. The opt-aiNET and N-opt-aiNET methodologies were successfully validated for a test system consisting of 13 thermal units whose incremental fuel cost function takes into account the valve-point loading effects.

Simulation results show that the proposed opt-aiNET approaches are applicable and effective in the solution of EDPs. In this context, the result reported here using N-opt-aiNET is comparatively lower than recent studies presented in the literature, except to the results using hybrid particle swarm with SQP proposed in [26].

The N-opt-aiNET has great potential to be further applied to many ill- conditioned problems in power system planning and operations. Additionally, we want to explore the benefits of using a belief space in other types of problems. In this context, the authors are interested in extending N-opt-aiNET for multiobjective environmental/economic power dispatch problems.

Acknowledgements

This work was supported by the National Council of Scientific and Technologic Development of Brazil — CNPq — under Grant 309646/2006-5/PQ.

References

1. Walters, D.C., Sheblè, G.B.: Genetic algorithm solution of economic dispatch with valve point loading. IEEE Trans. Power Systems 8(3), 1325–1332 (1993)
2. Lin, W.M., Cheng, F.S., Tsay, M.T.: Nonconvex economic dispatch by integrated artificial intelligence. IEEE Trans. Power Systems 16(2), 307–311 (2001)
3. Wood, A.J., Wollenberg, B.F.: Power generation, operation and control. John Wiley & Sons, New York (1994)
4. Lin, C.E., Viviani, G.L.: Hierarchical economic dispatch for piecewise quadratic cost functions. IEEE Trans. Power Apparatus and Systems 103(6), 1170–1175 (1984)
5. Liu, D., Cai, Y.: Taguchi method for solving the economic dispatch problem with nonsmooth cost functions. IEEE Trans. Power Systems 20(4), 2006–2014 (2005)
6. Victoire, T.A.A., Jeyakumar, A.E.: Reserve constrained dynamic dispatch of units with valve-point effects. IEEE Trans. Power Systems 20(3), 1273–1282 (2005)
7. Yang, H.T., Yang, P.C., Huang, C.L.: Evolutionary programming based economic dispatch for units with non-smooth fuel cost functions. IEEE Trans. Power Systems 11(1), 112–118 (1996)
8. Yalcinoz, T., Altun, H.: Power economic dispatch using a hybrid genetic algorithm. IEEE Power Engineering Review 21(3), 59–60 (2001)
9. Song, Y.H., Chou, C.S.V.: Large-scale economic dispatch by artificial ant colony search algorithms. Electric Machines and Power Systems 27(7), 679–690 (1999)
10. Park, J.-B., Lee, K.-S., Shin, J.-R., Lee, K.Y.: A particle swarm optimization for economic dispatch with nonsmooth cost function. IEEE Trans. Power Systems 20(1), 34–42 (2005)
11. Coelho, L.S., Mariani, V.C.: Combining of chaotic differential evolution and quadratic programming for economic dispatch optimization with valve-point effect. IEEE Trans. Power Systems 21(2), 989–996 (2006)
12. Papageorgiou, L.G., Fraga, E.S.: A mixed integer quadratic programming formulation for the economic dispatch of generators with prohibited operating zones. Electric Power Systems Research 77(10), 1292–1296 (2007)
13. Dasgupta, D.: Advances in artificial immune systems. IEEE Computational Intelligence Magazine 1(4), 40–49 (2006)
14. Castro, L.N., Timmis, J.: Artificial immune systems: a new computational intelligence approach. Springer, London (2002)
15. Castro, L.N., Von Zuben, F.J.: AINET: an artificial immune network for data analysis. In: Abbas, H., Sarker, R., Newton, C. (eds.) Data Mining: a heuristic approach. Idea Group Publishing (2001)
16. Castro, L.N., Timmis, J.: An artificial immune network for multimodal function optimization. In: Proceedings of IEEE Congress on Evolutionary Computation, Hawaii, HI, USA, pp. 699–674 (2002)
17. Sinha, N., Chakrabarti, R., Chattopadhyay, P.K.: Evolutionary programming techniques for economic load dispatch. IEEE Trans. Evolutionary Computation 7(1), 83–94 (2003)

18. Reynolds, R.: An adaptive computer model of the evolution of agriculture for hunter-gatherers in the valley of Oaxaca, Mexico, Ph.D. thesis, University of Michigan, Ann Arbor, MI, USA (1979)
19. Reynolds, R.: An introduction to cultural algorithms. In: Sebald, A.V., Fogel, L.J. (eds.) Proceedings of the 3rd Annual Conference on Evolutionary Programming, pp. 131–139. World Scientific Publishing, River Edge (1994)
20. Reynolds, R.G., Peng, B., Alomari, R.S.: Cultural evolution of ensemble learning for problem solving. In: Proceedings of IEEE Congress on Evolutionary Computation, Vancouver, BC, Canada, pp. 3864–3871 (2006)
21. Reynolds, R.G., Chung, C.: A self-adaptive approach to representation shifts in cultural algorithms. In: Proceedings of IEEE Congress on Evolutionary Computation, Nagoya, Japan, pp. 94–99 (1996)
22. Rychtyckyj, N., Reynolds, R.G.: Using cultural algorithms to improve knowledge base maintainability. In: Genetic and Evolutionary Computation Conference (GECCO 2001), San Francisco, CA, USA, pp. 1405–1412 (2001)
23. Coelho, L.S., Mariani, V.C.: An efficient particle swarm optimization approach based on cultural algorithm applied to mechanical design. In: Proceedings of IEEE Congress on Evolutionary Computation, Vancouver, BC, Canada, pp. 3844–3848 (2006)
24. Campelo, F., Guimarães, F.G., Igarashi, H., Ramírez, J.A., Noguchi, S.: A modified immune network algorithm for multimodal electromagnetic problems. IEEE Trans. Magnetics 42(4), 1111–1114 (2006)
25. Wong, K.P., Wong, Y.W.: Genetic and genetic/simulated-annealing approaches to economic dispatch. IEE Proc. Control, Generation, Transmission and Distribution 141(5), 507–513 (1994)
26. Victoire, T.A.A., Jeyakumar, A.E.: Hybrid PSO-SQP for economic dispatch with valve-point effect. Electric Power Systems Research 71(1), 51–59 (2004)

Scheduling of Tasks in Multiprocessor System Using Hybrid Genetic Algorithms

Betzy Varghes, Alamgir Hossain, and Keshav Dahal

Modeling Optimization Scheduling And Intelligent Control (MOSAIC) Research Centre
Department of Computing, University of Bradford, Bradford, BD7 1DP, UK
betzymol@yahoo.com, {m.a.hossain1,k.p.dahal}@bradford.ac.uk

Abstract. This paper presents an investigation into the optimal scheduling of real-time tasks of a multiprocessor system using hybrid genetic algorithms (GAs). A comparative study of heuristic approaches such as 'Earliest Deadline First (EDF)' and 'Shortest Computation Time First (SCTF)' and genetic algorithm is explored and demonstrated. The results of the simulation study using MATLAB is presented and discussed. Finally, conclusions are drawn from the results obtained that genetic algorithm can be used for scheduling of real-time tasks to meet deadlines, in turn to obtain high processor utilization.

Keywords: Optimal scheduling; hard real-time tasks; multiprocessor system; heuristics; genetic algorithm.

1 Introduction

Optimal scheduling is an important aspect in real-time systems to ensure soft/hard timing constraints. Scheduling tasks involves the allotment of resources and time to tasks, to satisfy certain performance needs [1]. In a real-time application, tasks are the basic executable entities that are scheduled [2]. The tasks may be periodic or aperiodic and may have soft or hard real-time constraints. Scheduling a task set consists of planning the order of execution of task requests so that the timing constraints are met. Multiprocessors have emerged as a powerful computing means for running real-time applications, especially where a uniprocessor system would not be sufficient enough to execute all the tasks by their deadlines [3]. The high performance and reliability of multiprocessors have made them a powerful computing means in time-critical applications [4]. In multiprocessor systems, the scheduling problem is to determine when and on which processor a given task executes.

Real-time task scheduling could be done either statically or dynamically. Dynamic schedule for a set of tasks is computed at run-time based on the tasks that is really executing. Static schedule on the other hand is done at compile time for all possible tasks. In the case of preemptive scheduling, an executing task may be pre-empted and the processor allocated to a task with higher priority or a more urgent task [2].

Real-time systems make use of scheduling algorithms to maximize the number of real-time tasks that can be processed without violating timing constraints [5]. A scheduling algorithm provides a schedule for a task set that assigns tasks to processors

E. Avineri et al. (Eds.): Applications of Soft Computing, ASC 52, pp. 65–74.
springerlink.com   © Springer-Verlag Berlin Heidelberg 2009

and provides an ordered list of tasks. The schedule is said to be feasible if the timing constraints of all the tasks are met [2]. All scheduling algorithms face the challenge of creating a feasible schedule.

A number of algorithms have been proposed for dynamic scheduling of real-time tasks. It is said that there does not exist an algorithm for optimally scheduling dynamically arriving tasks with or without mutual exclusion constraints [6]. This has motivated the need for heuristic approaches for solving the scheduling problem. Page and Naughton [7] gives a number of references in which artificial intelligence techniques are applied for task scheduling. They have also reported good results from the use of GAs in task scheduling algorithms.

This paper aims to provide an insight into scheduling real-time tasks by using genetic algorithm incorporating traditional scheduling heuristics to generate a feasible schedule based on the work done by Mahmood [5]. That is, the use of a hybrid genetic algorithm to dynamically schedule real-time tasks in multiprocessor systems. The scheduling algorithm considered, aims in meeting deadlines and achieving high utilization of processors. The paper also provides a comparative study of on applications of heuristic approaches, such as 'EDF' and 'SCTF' separately, and genetic algorithms. The scheduler model considered for the study would contain task queues from which tasks would be assigned to processors. Task queues of varying length would be generated at run time. From the task queue only a set of tasks would be considered at a time for scheduling. The size of the task sets considered for scheduling would also be varied for a comparative study. The MATLAB software tool was used for the simulation study as it integrates computation, visualization and programming in an easy to use environment.

2   Related Research Work

Scheduling algorithms for multiprocessor real-time systems are significantly more complex than the algorithms for uniprocessor systems [5]. In multiprocessor systems, the scheduling algorithm, other than specifying the ordering of tasks must also determine the specific processors to be used.

Goossens and others have been studying the scheduling of real-time systems using EDF scheduling upon uniform and identical multiprocessor platforms. They justify that EDF remains a good algorithm to use in multiprocessor systems. They also propose a new priority-driven scheduling algorithm for scheduling periodic task systems upon identical multiprocessors [8].

Manimaran and Siva Ram Murthy [4] say that there does not exist an algorithm for optimally scheduling dynamically arriving tasks with or without mutual exclusion constraints on a multiprocessor system. This has stimulated the need for developing heuristic approaches for solving the scheduling problem. In heuristic scheduling algorithms, the heuristic function H evaluates various characteristics of tasks and acts as a decision aid for the real-time scheduling of the tasks.

Myopic scheduling algorithm is another heuristic scheduling algorithm for multiprocessor systems with resource constrained tasks. The algorithm selects a suitable process based on a heuristic function from a subset; referred to as window, of all ready processes instead of choosing from all available processes like the original heuristic scheduling algorithm. Another difference of the algorithm from the original heuristic algorithm is that the unscheduled tasks in the task set are always kept sorted by increasing order of deadlines. Hasan et al. [9] presents the impact of the performance in implementing the myopic algorithm for different window sizes.

Page and Naughton [7] gives a number of references to examples where artificial intelligence techniques are being applied to task scheduling. They say that techniques such as genetic algorithms are most applicable to the task scheduling problem because of the need to quickly search for a near optimal schedule out of all possible schedules. The paper presents a scheduling strategy which makes use of a genetic algorithm to dynamically schedule heterogeneous tasks on heterogeneous processors in a distributed system [7]. Genetic algorithm has been utilized to minimize the total execution time. The simulation studies presented shows the efficiency of the scheduler compared to a number of other schedulers. However the efficiency of the algorithm for time critical applications has not been studied.

Oh and Wu [10] presents a method for real-time task scheduling in multiprocessor systems with multiple objectives. The objectives are to minimize the number of processors required and also minimize the total tardiness of the tasks. A multi-objective genetic algorithm has been made use of for scheduling to achieve optimization. The work considers scheduling tasks of precedence and timing constrained task graph. The algorithm was shown to give good performance. While Oh and Wu [10] focuses on multiobjective optimization, the algorithm discussed in the paper aims to meeting deadlines of tasks and achieving high resource utilization. There are also examples where genetic algorithm has been used for scheduling tasks in uniprocessor systems. Yoo and Gen [11] presents a scheduling algorithm for soft real-time systems. They have used proportion-based genetic algorithm and focused mainly on the scheduling of continuous tasks that are periodic and preemptive.

The scheduling algorithm presented in this paper is based on the work done by Mahmood [5]. The genetic algorithms in their purest form could be called as blind procedures. They do not make use of any problem specific knowledge which may speed up the search or which may lead to a better solution. That is why specialized techniques which make use of problem specific knowledge out-performs genetic algorithms in both speed and accuracy. Therefore it may be advantageous to exploit the global perspective of the genetic algorithm and the convergence of the problem specific techniques.

3   System Model

As discussed earlier, dynamically scheduling tasks in a multiprocessor system using a hybrid genetic algorithm presented in the following sections is based on the principle of the work done by Mahmood [5]. The task and scheduler model for the simulation system considered is discussed below.

Task Model: The real-time system is assumed to consist of m, where $m > 1$, identical processors for the execution of the scheduled tasks. They are assumed to be connected through a shared medium. The scheduler may assign a task to any one of the processors. Each task T_i in the task set is considered to be aperiodic, independent and nonpreemptive.

Each task T_i is characterised by: A_i : arrival time; R_i : ready time; C_i : worst case computation time; D_i : deadline.

The scheduler determines the scheduled start time and finish time of a task. If $st(T_i)$ is the scheduled start time and $ft(T_i)$ is the scheduled finish time of task T_i, then the task T_i is said to meet its deadline if $(R_i \leq st(T_i) \leq D_i - C_i)$ and $(R_i + C_i \leq ft(T_i) \leq D_i)$. That is, the tasks are scheduled to start after they arrive and finish execution before their deadlines [3]. A set of such tasks can be said to be guaranteed.

Scheduler Model: As discussed before the dynamic scheduling in a multiprocessor system could be either centralized or distributed. This paper assumes a centralized scheduling scheme with each processor executing the tasks that fill its dispatch queue. Since a centralized scheduling scheme is considered, all the tasks arrive at a central processor called the scheduler. The scheduler has a task queue associated with it to hold the newly arriving tasks. Thus the incoming tasks are held in the task queue and then passed on to the scheduler for scheduling of tasks. It is the central scheduler that allocates the incoming tasks to other processors in the system.

Each processor has a dispatch queue associated with it. The processor executes tasks in the order they arrive in the dispatch queue. The communication between the scheduler and the processors is through these dispatch queues. The scheduler works in parallel with the processors. The scheduler schedules the newly arriving tasks and updates the dispatch queue while the processors execute the tasks assigned to them. The scheduler makes sure that the dispatch queues of the processors are filled with a minimum number of tasks so that the processors will always have some tasks to execute after they have finished with their current tasks. Thus the processing power can be utilized without making it idle.

The minimum capacity of the dispatch queues depends on factors like the worst case time complexity of the scheduler to schedule newly arriving tasks [6]. A feasible schedule is determined by the scheduler based on the worst case computation time of tasks satisfying their timing constraints.

The scheduling algorithm to be discussed has full knowledge about the set of tasks that are currently active. But it does not have knowledge about the new tasks that may arrive while scheduling the current task set.

The objective of the dynamic scheduling is to improve or maximize what is called the guarantee ratio. It is defined as the percentage of tasks arrived in the system whose deadlines are met. The scheduler in the system must also guarantee that the tasks already scheduled will meet their deadlines.

4   The Scheduling Algorithm

A hybrid genetic algorithm for scheduling real-time tasks in multiprocessor system is discussed in this section. Initially a task queue is generated with tasks having the following characteristics namely, arrival time, ready time, worst case computation time and deadline. The tasks are sorted in the increasing order of their deadlines. The tasks are ordered so that the task with the earliest deadline can be considered first for scheduling. The algorithm considers a set of tasks from the sorted list to generate an initial population. In the initial population, each chromosome is generated by assigning each task in the task set to a randomly selected processor and the pair (task, processor) is inserted in a randomly selected unoccupied locus of the chromosome. The length of the chromosome depends on the number of tasks selected from the sorted list. The tasks in each chromosome are then sorted based on their deadline. This is done because the chromosome representation also gives the order in which the tasks are executed in a processor. The sorting ensures that the tasks with earliest deadline are given priority. The fitness evaluation of the chromosomes in the population is then performed. The fitness value of a chromosome is the number of tasks in the chromosome that can meet their deadlines (i.e. the objective is to maximize the number of tasks in each chromosome that meet their deadlines). The chromosomes in the population are then sorted in the descending order of their fitness value.

Genetic operators are then applied to the population of chromosomes until a maximum number of iterations have been completed. When applying genetic operators to the population, selection is applied first followed by crossover, partial-gene mutation, sublist-based mutation and then order-based mutation. In each iteration, the tasks in the chromosomes are sorted based on their deadline and the evaluation of the chromosomes and sorting of the chromosomes based on fitness value is performed. After number of iterations the best schedule for the set of tasks is obtained.

The tasks that are found infeasible are removed from the chromosomes so that they are not reconsidered for scheduling. For a task T_i to be feasible it should satisfy the condition that $(R_i \leq st(T_i) \leq D_i - C_i)$ and $(R_i + C_i \leq ft(T_i) \leq D_i)$ where R_i is the ready time, D_i is the deadline and C_i is the worst case computation time of task T_i. $st(T_i)$ and $ft(T_i)$ denoted the start time and finish time of task T_i respectively. If the condition is not satisfied it is said to be infeasible.

5   Implementation and Results

The simulation study (using MATLAB) considers the assigning of a set of tasks to a number of processors. For these, task queues of different lengths were generated at run time from which a set of tasks were chosen at a time for scheduling. The lengths of task queues considered were 100, 200, 400 and 600. The worst case computation time, C_i, of a task T_i has been chosen randomly between a minimum and maximum computation time value denoted by MIN_C and MAX_C. The values of MIN_C and MAX_C were set to 30 and 60 respectively. The value for the deadline of a task T_i has been randomly chosen between $(R_i + 2 * C_i)$ and $(R_i + r * C_i)$ where $r \geq 2$. This ensures that the computation time is always less than the deadline. For the study, the value of r has been chosen to be 3. The mean of the arrival time was assumed to be 0.05. The number of processors, m considered was 10.

The values for the number of iterations for the application of the genetic operators have been based on number of trials. For the value of 'x', which denotes the percentage of tasks to be killed before applying reproduction operator, it has been reported in [5] that best results were obtained with x = 20. Therefore the value of 20 percent has been considered for the algorithm presented in the paper. The chromosome size has been assumed equal to the number of tasks considered at a time for scheduling. Depending on this, the value for the chromosome size has been varied between 20 and 60. As mentioned before the fitness value determines the number of tasks in the chromosome that can meet their deadlines, i.e., the number of tasks that are feasible.

Hence here, for chromosome size 20 the maximum fitness value that can be obtained is 20. The population size for the algorithm has been assumed to be 30. That is 30 chromosomes have been considered at a time for the application of genetic operators. Thus the tasks which have been generated with the values for their characteristics chosen appropriately have been considered for scheduling. Initially the tasks were assigned to processors based on 'Earliest Deadline First'. After the results have been observed, the tasks were scheduled using the proposed hybrid genetic algorithm. The algorithm was then implemented by incorporating the heuristic 'Shortest Computation time First' with genetic algorithm. Set of tasks were scheduled using the modified algorithm and the results were observed.

For an initial evaluation the fitness value by assigning tasks based on Earliest Deadline First (EDF) was calculated. For this, a task queue of 100 tasks was generated randomly and it was divided into task sets of 20 each. The tasks were ordered in the increasing order of their deadlines and assigned to processors considering earliest deadline first. The processors were chosen randomly between 1 and 10. The fitness value obtained for each task set is shown in Figure 1. The graph shows that the maximum number of tasks that meet their deadlines is 16 when considering 20 tasks for scheduling. The majority of the task sets gave a fitness value of 12.

The hybrid algorithm presented in the paper was then used to schedule the same task sets. The algorithm incorporates the heuristic 'Earliest Deadline First' and also genetic algorithm. Here also a set of 20 tasks was considered at a time. The graph showing the fitness value of tasks obtained using the algorithm is shown in Figure 2.

As shown by the graph, a better performance is obtained by using genetic algorithm with the heuristic. Thus it could be seen that, the percentage of tasks that are feasible is 95 percent and above. The algorithm was also studied for different task sets with the same chromosome size. In all the cases the percentage of tasks that are feasible was always 90 percent and above when the chromosome size considered was 20. These demonstrate that genetic algorithm could be used to schedule task to meet deadlines

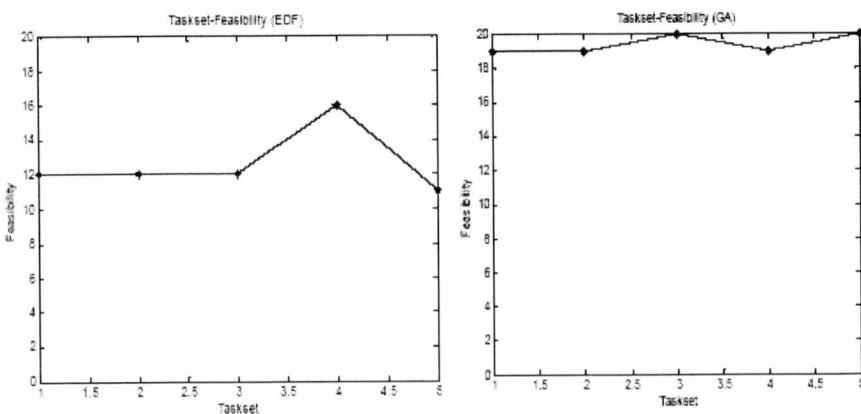

Fig. 1. Task set feasibility based on EDF   **Fig. 2.** Task set feasibility using hybrid GA

and also achieve better processor utilization. However, it is worth noting that genetic algorithms do have the disadvantage of spending much time in scheduling.

As mentioned earlier in the paper, the population size for the genetic algorithm was taken to be 30. In the initial population the fitness value of chromosomes were low. As the number of iterations increases a better solution is obtained. The number of iterations considered for the algorithm was 50.

A graph which depicts the change in the feasibility value from the initial to the final iteration for a particular task set of 20 tasks is shown in Figure 3. The graph shows that the fitness value of chromosomes changes gradually from a minimum value of 12 to a maximum value of 20. Thus a better solution can be obtained by applying genetic algorithm for a good number of iterations. The number of iterations needed for the genetic operators was based on a trial method. This was mainly considered for the chromosome size 20.

The results of incorporating the heuristic 'Earliest Deadline First' with genetic algorithm demonstrated better performance. This motivated to study the efficiency of the algorithm by incorporating other heuristics. The heuristic, Shortest Computation time First (SCF) was incorporated with genetic algorithm for this.

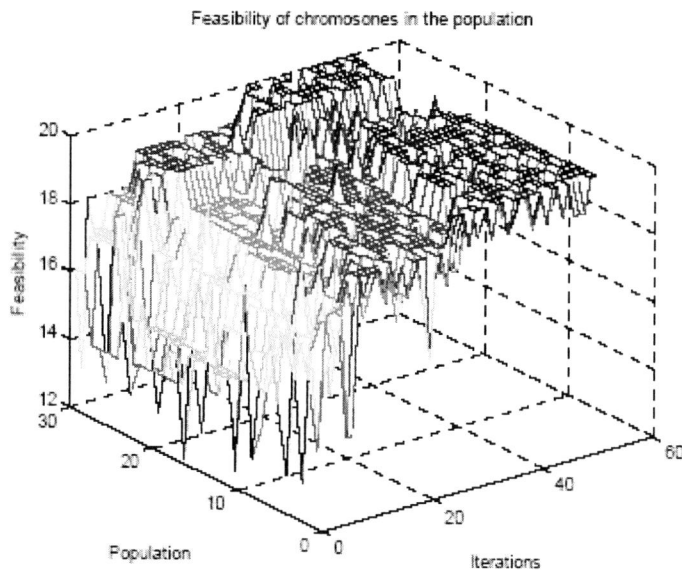

Fig. 3. Feasibility value vs change of chromosomes in the population

For the study, the chromosome size was kept at 20. The length of the task queue considered was 100, like before. The algorithm was slightly modified to incorporate SCF heuristic. In the case where the tasks were sorted based on the deadline, the algorithm was modified so that the tasks were sorted based on their computation time. The tasks were sorted in the increasing order of computation time. The fitness function was not changed. It determines the number of tasks that can be scheduled without missing their deadline. It was seen that, the result was almost similar to that obtained in the case of using earliest deadline first. That is to say, it gave almost similar performance.

It was then decided to change the length of the task queue while maintaining the chromosome size at 20 and the not altering anything else. The results were compared for the two cases, that is, using earliest deadline first and shortest computation time first. The task queue lengths considered were 100, 200, 400 and 600. The comparison of the heuristics has been made based on the fitness value. As the chromosome size has been fixed at 20, the maximum value for fitness that can be obtained is 20. It could be seen that for all the cases the number of tasks that were feasible was 90 percent and above for both the heuristics. This gives the impression that the heuristic shortest computation first could also be incorporated with genetic algorithm to give feasible solutions. The graph of the comparison is shown in the Figure 4. This demonstrates a better overview of the results discussed above.

The results were then compared for task queues of different length by changing the chromosome size. The lengths of task queue considered were same as before namely, 100, 200, 400 and 600. The chromosome size chosen were 40 and 60. Though both the heuristics showed almost similar performance in the case of chromosome size 20, the result was not same for higher values of chromosome size. It could be seen that the use of heuristic shortest computation time first gave better fitness values compared to earliest deadline first when incorporated with genetic algorithm. This shows that the heuristic shortest computation time first is a better option for incorporating with genetic algorithm. Fig. 5 shows the comparison of the heuristics based on fitness value for chromosome size 40.

Table 1 shows the comparative fitness function of SCF and EDF. It is noted that only 48 percent of the tasks could be scheduled when the chromosome size is 60, whereas in the case with chromosome size 20, nearly 100 percent of the tasks could be scheduled. It should be mentioned that the result considers a fixed number of

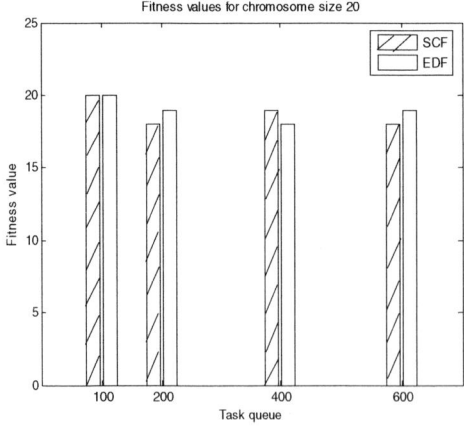
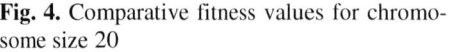
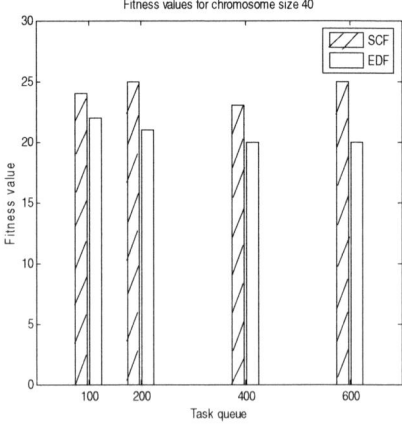

Fig. 4. Comparative fitness values for chromosome size 20

Fig. 5. Comparative fitness values for chromosome size 40

processors, i.e. 10. Thus a comparative study shows that best results are obtained with chromosome size 20. It could also be noted that better results are obtained when the length of the task queue is 100.

From the above results it could be said that traditional scheduling heuristics could be incorporated with genetic algorithm to schedule real-time tasks if the scheduling time used by genetic algorithm is reduced by some efficient method.

Table 1. Fitness value obtained for different Chromosome size

Task queue	Fitness value			
	Chromosome size 20		Chromosome size 40	
	SCF	EDF	SCF	EDF
100	20	20	24	22
200	18	19	25	21
400	19	18	23	20
600	18	19	25	20

6 Conclusion

A hybrid genetic algorithm for scheduling tasks in multiprocessor system has been presented based on the work done by Mahmood [5]. The paper has discussed that genetic algorithm incorporating traditional heuristics could be used to obtain optimal solutions. A comparative performance of using heuristics EDF and SCTF with genetic algorithm has been presented and discussed through a set of experiments. It is noted that incorporating SCTF with genetic algorithm offered better performance as compared to the EDF. The algorithm presented in the paper has been successful in obtaining feasible solutions for a task set of 20 and also achieving high utilization of processors.

However it is noted that the implementation of the genetic algorithm is quite costly since populations of solutions are coupled with computation intensive fitness evaluations. This can be overcome by employing high performance computing platform or parallel processing technique in multiprocessor computing domain.

References

[1] Ramamritham, K., Stankovic, J.A.: Scheduling Algorithms and Operating Systems Support for Real-time Systems. Proceedings of IEEE 82(1), 55–67 (1994)
[2] Cottet, F., Delacroix, J., Kaiser, C., Mammeri, Z.: Scheduling in Real-time Systems, pp. 1–64. John Wiley & Sons Ltd., England (2002)
[3] Eggers, E.: Dynamic Scheduling Algorithms in Real-time, Multiprocessor Systems, Term paper 1998-1999, EECS Department, Milwaukee School of Engineering, North Broadway, Mil-waukee, WI, USA (January 1999)

[4] Manimaran, G., Siva Ram Murthy, C.: An Efficient Dynamic Scheduling Algorithm for Multiprocessor Real-time Systems. IEEE Transactions on Parallel and Distributed Systems 9(3), 312–319 (1998)

[5] Mahmood, A.: A Hybrid Genetic Algorithm for Task Scheduling in Multiprocessor Real-Time Systems. Journal of Studies in Informatics and Control 9(3) (2000) (accessed on 27/06/2005), http://www.ici.ro/

[6] Manimaran, G., Siva Ram Murthy, C.: A Fault-tolerant Dynamic Scheduling Algorithm for Multiprocessor Real-time Systems and Its Analysis. IEEE Transactions on Parallel and Dis-tributed Systems 9(11), 1137–1152 (1998)

[7] Page, A.J., Naughton, T.J.: Dynamic task scheduling using genetic algorithms for heterogeneous distributed computing. In: 8th International Workshop on Nature Inspired Distributed Computing, proceedings of the 19th International Parallel & Distributed Processing Symposium, Denver, Colorado, USA. IEEE Computer Society, Los Alamitos (2005)

[8] Goossens, J., Baruah, S., Funk, S.: Real-time Scheduling on Multiprocessors (2002) (accessed on 12/08/05), http://citeseer.ist.psu.edu/

[9] Hasan, M.S., Muheimin-Us-Sak, K., Hossain, M.A.: Hard Real-Time Constraints in Implementing the Myopic Scheduling Algorithm. International Journal of High Performance Computing Applications (to appear, 2005)

[10] Oh, J., Wu, C.: Genetic-algorithm-based real-time task scheduling with multiple goals. The Journal of Systems and Software 71(3), 245–258 (2004)

[11] Yoo, M.R., Gen, M.: Bicriteria real-time tasks scheduling using proportion-based genetic algorithm, August 15, 2005, pp. 213–222 (2001),
http://www.complexity.org.au/conference/upload/yoo01/yoo01.pdf

A Novel Modelling and Optimisation Technique for Business Processes: An Evolutionary Computing Based Approach

Mark Stelling, Rajkumar Roy, and Ashutosh Tiwari

Decision Engineering Centre, Cranfield University, Cranfield, MK43 0AL, UK

Summary. The ideas discussed in this paper are based on the research question "how can a business process be represented as a chromosome?". By constructing a company taxonomy and attaching to it a classification and coding system, a process model can be comprehensively represented as a numerical chromosome with the aim of then optimising it via the use of an evolutionary computation based technique. This paper suggests a methodology for analysing and coding a process, and goes on to demonstrate how the process is optimised, in this instance based on cost evaluation only. All research conducted has concentrated on the service sector and this type of business process is used as an example.

1 Introduction

A business process can be described as "*a set of activities, which can be broken down into tasks, that when taken together take an input, transform it, and produce an output*" [1]. All organisations use processes - they may not be formalised and represented as a diagram or model but they will always exist, no matter what type of business the organisation is involved in. There has always been the desire to make these processes as efficient as possible — why would a company consistently want to waste resources when their processes could be improved [2]? This research considers process improvement based on cost. A means of preparing a process for optimisation is demonstrated followed by the actual optimisation procedure and results using a hybrid technique incorporating both rule-based programming and a genetic algorithm. This paper follows on from the work described in the authors previous conference paper, "*A Coding Mechanism for Business Process Optimisation Using Genetic Algorithms: Initial Thoughts*" [2], in which the technique described here was presented in its initial stages of development.

2 Background

2.1 Process Modelling

The representation of a business process by means of a model allows a common and comprehensive understanding of the process and enables analysis of the

E. Avineri et al. (Eds.): Applications of Soft Computing, ASC 52, pp. 75–85.
springerlink.com © Springer-Verlag Berlin Heidelberg 2009

process to be carried out. There are a large number of modelling techniques and tools, some of which are more widely used in industry than others. These include UML modelling language (object oriented activity diagrams), IDEF (Integrated Definition for Function Modelling), Petri Nets [3], [4] and modelling standard BPMN (Business Process Modelling Notation). BPMN aims to provide a notation that is readily understandable by all business users — its intent being to standardise business process modelling notation [5]. All process representations in this research project adhere to BPMN standards, along with the terminologies used to describe all elements of the model.

2.2   Business Process Optimisation

There are very few optimisation techniques to be found in literature which are suitable for application to business processes. This is primarily due to the qualitative nature of many process representations which makes analysis difficult [6]. As a result, there is a lack of algorithmic approaches to the optimisation of business processes [7].

Zakarian succeeded in quantifying the variables in an IDEF3 process model using IF-THEN fuzzy rules and was then able to estimate the output from the process [8]. A similar approach was followed by Phalp when he tried to quantify Role Activity Diagrams, thus providing data for optimisation [9]. One method which has some degree of relevance to the ideas proposed in this paper involved the numerical representation of a process design in the form of a three-dimensional cube consisting of 5 * 5 * 5 binary chromosomes depicting activities and resources [10].

2.3   Classification and Coding Systems

Many organisations have developed their own classification and coding systems. This applies not only to the manufacturing and production industries who use classification and coding as a part of their group technology structure, but also to many record keeping organisations who hold vast numbers of diverse records.

According to the National Archives of Australia, the structure of a classification system is usually hierarchical and reflects the analytical process as follows:

- first level — the business **function**
- second level — **activities** constituting the function
- third level — groups of **transactions** that take place within each activity [11]

Many different types of coding systems exist in industry. Most will share similar characteristics and follow similar principles. These principles include the following:

- The shorter the code, the fewer the errors
- Identifying codes should be of fixed length and pattern
- Purely numeric codes produce less errors than alphanumeric combinations, but these are acceptable if the alphabetic field is fixed and used to break a string of digits

Although purely numeric systems are less prone to error, many coding systems use a combination of alphabetic and numeric characters - for instance, FA02 could be used for Finance and Accounting department, business expenses section. This would make them unsuitable for the process representation suggested later in this paper and so it was necessary for the author to invent a purely numerical coding system based on some of the ideas gained by research into many of the coding methods currently in use.

2.4   Taxonomies

Taxonomies are a method of classification and play a fundamental role in the system of representing a business process as a numerical chromosome.

The word taxonomy, associated principally with biology, is defined as *"the science of finding, describing, classifying and naming organisms"*, *"the branch of science concerned with classification"*, and also as *"a scheme of classification"* [12], [13]. Taxonomies are often hierarchical structures consisting of parent-child relationships but may also refer to relationship schemes other than hierarchies, such as network structures.

Taxonomies can be applied to any business, whether it be to the structure of the company personnel, the classification of its records or the organization of its business processes - by breaking down the processes into hierarchical structures beginning with the root node, 'domain', underneath which may be multiple categories of 'activities' followed by 'transactions' or 'topics'.

This tree structure can be extended further to include sub-levels such as 'sub-domain', 'sub-transaction', 'sub-sub-sub transaction' and so on, until an acceptable degree of detail is reached. This degree of detail and the naming of the levels are dependent on the intended purpose of the taxonomy. It is proposed in this paper to allow for differing levels of domain and transaction: these levels will be described later in the paper.

The purpose of using taxonomies in this project is to allow the organisation of a companys business processes to be depicted and subsequently assigned codes in order to represent the processes in a purely numeric way.

3   Coding Mechanism

This project requires a very specific, multi-level taxonomy in order to allow the assignment of codes to the process instances at the bottom of the hierarchy. This entails an extensive breakdown of a business structure from the highest level down to the lowest level of process instances. A case study of a service based company was conducted and a company taxonomy was designed, based on the data gathered during the study. The example in Figure 1 shows one section of the taxonomy, detailing only the customer services/contact/orders/place telephone order levels of the structure.

It is necessary to break down this taxonomy into one more level of detail ('sub-transaction' or 'Transaction — Level 2') in order to fully represent a process.

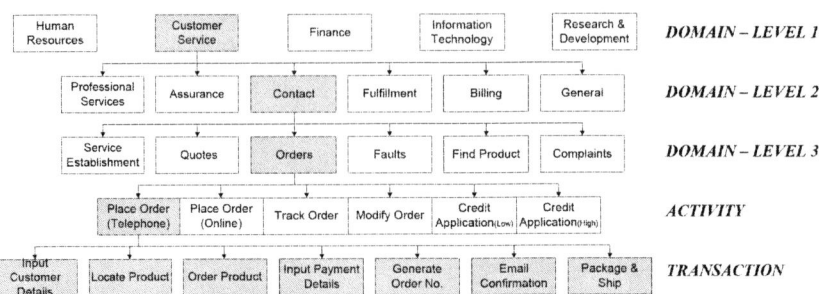

Fig. 1. Service Sector Taxonomy

Table 1 shows a simple, purely numeric taxonomy coding system which appears suitable for the aims of the project.

In this example, the code numbers are all sequential and start from 1. This may not be as straight forward had the whole company taxonomy been coded.

Table 1. Taxonomy Coding System

CODE	LEVEL	NAME
1	DOMAIN L1	HUMAN RESOURCES
2	DOMAIN L1	CUSTOMER SERVICE
3	DOMAIN L1	FINANCE
4	DOMAIN L1	INFORMATION TECHNOLOGY
1	DOMAIN L2	PROFESSIONAL SERVICES
2	DOMAIN L2	ASSURANCE
3	DOMAIN L2	CONTACT
4	DOMAIN L2	FULFILMENT
5	DOMAIN L2	BILLING
6	DOMAIN L2	GENERAL
01	DOMAIN L3	SERVICE ESTABLISHMENT
02	DOMAIN L3	QUOTES
03	DOMAIN L3	ORDERS
04	DOMAIN L3	FAULTS
05	DOMAIN L3	FIND PRODUCT
06	DOMAIN L3	COMPLAINTS
01	ACTIVITY	PLACE ORDER (TELEPHONE)
02	ACTIVITY	PLACE ORDER (ONLINE)
03	ACTIVITY	TRACK ORDER
04	ACTIVITY	MODIFY ORDER
05	ACTIVITY	CREDIT APPLICATION (LOW)
06	ACTIVITY	CREDIT APPLICATION (HIGH)
01	TRANSACTION	RETRIEVE CUSTOMER DETAILS
02	TRANSACTION	RETRIEVE ORDER DETAILS
03	TRANSACTION	EMAIL ORDER STATUS

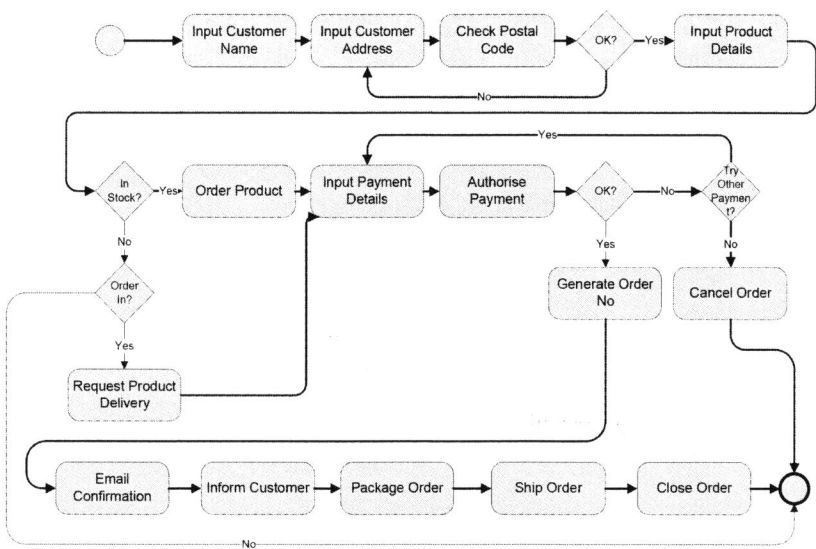

Fig. 2. Place Telephone Order Process

Now that the taxonomy has been assigned the necessary codes to identify its component domains, activities and transactions, it is necessary to construct a process flowchart from the codes used. The steps of the process have already been defined to a large degree in the taxonomy, but it is not possible to depict such elements as loops, decisions, branches etc. in the taxonomy structure — for instance, the transactions that constitute the above activity need to include many of these elements. The actual BPMN process model is more complex and would look like Figure 2.

Extra considerations are required to fully represent the complete process. Although the order of the process and its steps are exactly as in the taxonomy definition, extra details such as checking the customer address via their postcode, ascertaining whether or not the product is in stock and whether the payment is authorised are necessary. As such, these extra elements in the process also require codes so they can be included in the numerical process representation. These low level steps of the process are classed here as sub-transactions. The numerical representation of each element of a process is listed in Table 2 with some examples and definitions.

This mechanism gives a total of 174 digits per process instance. The codes Input, Output, Constraint, Resource, Transition Out and Step Type are crucial to the eventual evaluation of each element of the process. They have a range of possible values, allocated for the purposes of the example process presented in this paper - they will be expanded on in future development work. As an example, the start node of the activity, 'Place Telephone Order', shown in Figure 3, is represented as follows:

Table 2. Code Details

CODE NAME	NO. DIGITS	EXAMPLE/EXPLANATION
Domain Level 1	1	1=Human Resources
Domain Level 2	1	5=Billing
Domain Level 3	2	04=Faults
Activity	2	03=Track Order
Transaction	2	03=Order Product
Sub-Transaction	2	03=Check Postal Code
Input	15	Up to 3 possible alternative inputs of 5 digits E.G. 002050020600000=Placed Order or Placed Delivery
Output	15	Up to 3 possible alternative outputs of 5 digits E.G. 002030020400000=Pending Order or Pending Delivery
Constraint	9	Up to 3 constraints of 3 digits E.G. 001003000=Must be first step AND must be linked to next step
Resource	3	001=Internal Customer RDB
Flows In	10	Up to 5 incoming flows of 2 digits per flow (normally 1 except for join (AND-Join) & merging (OR-Join) process steps). E.G. 0101000000=2 incoming normal flows
Flows Out	10	Up to 5 outgoing flows of 2 digits per flow (normally 1 except for fork (AND-Split) & decision (OR-Split) process steps) E.G. 0109010000=1 outgoing normal flow, 1 sequence flow loop, 1 outgoing normal flow
Previous Step	50	Allows for up to 5 steps of 10 digits (Domain L1, Domain L2, Domain L3, Activity, Transaction, Sub-transaction)
Next Step	50	Allows for up to 5 steps of 10 digits (Domain L1, Domain L2, Domain L3, Activity, Transaction, Sub-transaction)
Step Type	2	20=Gateway — AND-Join

23030100000100000000000000010000010100010800100000000000000000000001000000000000000000000000000000
00000000000000000000000000000000230301010100000000000000000000000000000000000000000001

Fig. 3. Place Telephone Order - Start Node Code

The entire process translates into a code (or chromosome) of 3828 digits (22 process steps or genes). This string of digits can be input to a GA, as GA work with populations of chromosomes — these are often binary but not in this case.

4 Optimisation Technique

All process data is held in a relational database, in this case a Microsoft Access RDB.

The schema reflects the taxonomy structure, with extra lookup tables and tables containing cost details. Within the database there are a number of processes, or activities, made up of a number of process steps, or transactions and

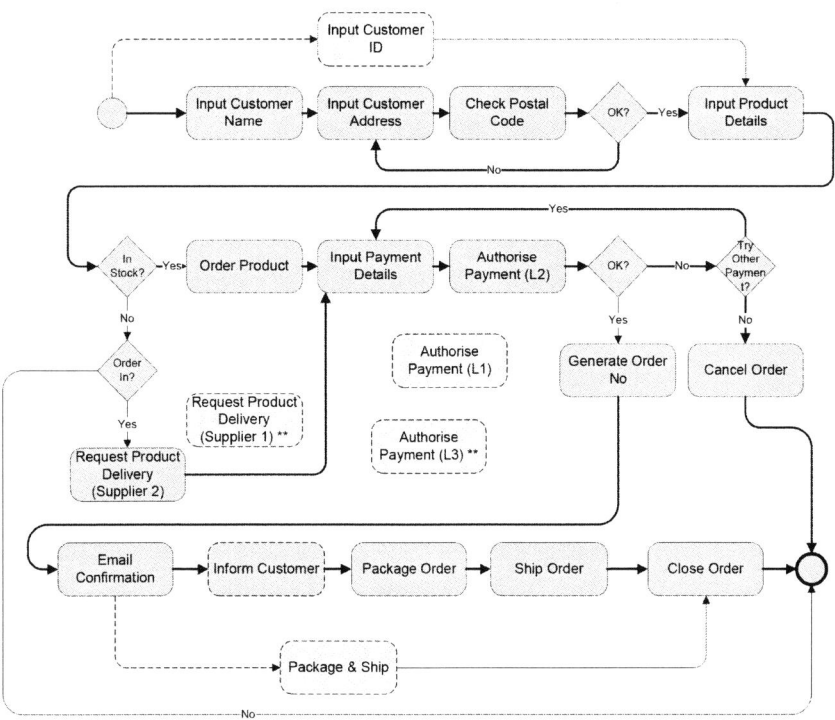

Fig. 4. Process Alternatives

sub-transactions. Subject to constraints being adhered to, all steps are inter-changeable - they could be used as a part of any process. There are also a number of steps available in the database that may not be used as a part of a process, but, if used to replace parts of an existing process, may lead to an optimal process (in terms of lower cost). The reason for using GA as a part of the optimisation technique is for its ability to generate vast numbers of poten-tial solutions — in this case, vast numbers of different processes. If there were 2 alternative task steps for each of the 15 existing tasks in the example tele-phone order process (figure 2), then the number of different processes that could be generated without shortening or reconfiguring the process would add up to 14,348,907 (i.e. 3^{15}). As all alternatives will have differing costs, the human effort involved in finding the optimum process would be considerable — thus the need for automation. The diagram in Figure 4 shows the example telephone order process from Figure 2 - some alternative replacement steps are available (one for one replacement) plus some single steps that will replace more than one existing step (these are the white boxes) and also one duplicate step (Inform Customer is exactly the same input/output as Email Confirmation). The number of potential process combinations available from the alternatives in this example is small and is purely for demonstrating the workings of the technique.

The optimisation technique will use these alternatives to re-engineer the process. The program is a combinatorial, hybrid algorithm incorporating standard, rule-based programming and GA. It uses SQL for data retrieval and VBA for all other program code.

The following guidelines apply to the optimisation technique:

- The optimised process will not be longer than the present one: it will be the same length or shorter.
- For purposes of efficiency, only feasible processes will be created during run-time.
- At this stage of development, the GA module will not attempt to reconfigure the process.

The high level functionality of the program is illustrated in the flowchart in Figure 5:

- The first part of the program attempts to find duplicates in the process and remove them (matching input and output indicates a duplicate). It then tries to remove steps from the existing process by checking inputs/outputs/constraints - storing input from one process step, looking at

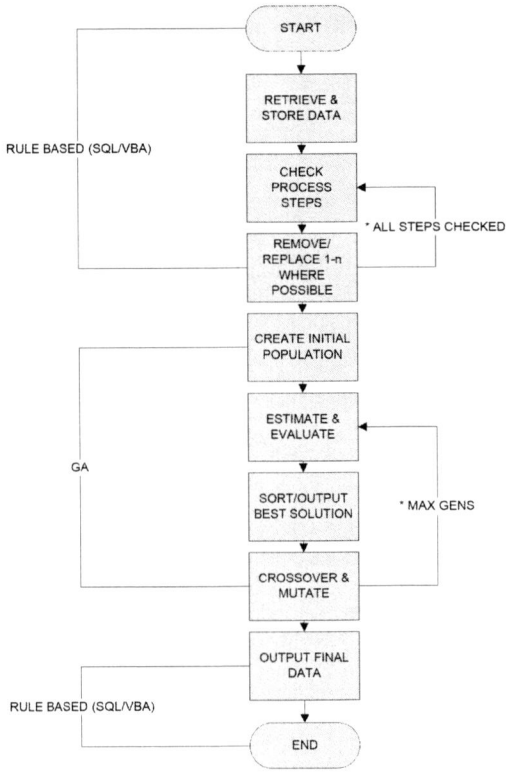

Fig. 5. Optimisation Program Flowchart

outputs later in the process and trying to find an alternative step with similar input to the one process step and similar output to a later process step — thus, subject to constraints not being broken, allowing the removal of unnecessary steps. If the program finds a suitable replacement for a number of steps, it replaces them with the alternative and then continues searching through the rest of the alternatives just in case an even better replacement is available.

- The initial population in the GA part of the program is created from all steps available in the search space, but only feasible processes are created: if no alternatives are available for an exiting process step, the original step is used; if one or more alternatives are available for a process step, they will be selected at random as replacements for that specific step (along with the original step).
- The GA will not attempt to reconfigure the process - it will attempt to find the optimum process by using all combinations of the existing process plus potential replacements. Uniform Crossover and mutation will take place and the loci will remain the same in all chromosomes (so as to maintain the feasibility of all solutions). In this way, the GA should evaluate every available process combination during its course and will only ever create feasible processes.

5   Results and Discussion

The first part of the program creates a shortened process. Due to the small number of alternative processes available (6) without removing any steps, the program always creates the optimum process in the second or third generation (out of 10 generations in this test) of the GA section. The results, in every run, are successful and create the process shown in Figures 6 and 7.

It should be noted that this example is a numerical one, used to demonstrate the workings of this technique, and the numbers are normalised - the process costs are a number of 'cost units' and are not intended to accurately portray the real costs.

The 'cost' of the present process is 212. This is reduced to 200 by removing the duplicate, then 150 by removing/replacing steps, and then to 135 by the GA section.

At present, the technique only looks for 'one for one' or 'many for one' replacements - it is intended to add a 'Feature Library' facility to the database which will provide potential replacements for whole sections of processes (i.e. a 'many for many' replacement facility). It is also intended to include time and customer satisfaction as evaluation criteria, making this a multi-objective optimisation technique.

There are implementation issues associated with this technique but, once a companys process models are assigned the necessary numerical codes and stored in the RDB, which is quite a time consuming process, it is simply a matter of periodic maintenance (i.e. when a new process step becomes available, it is coded

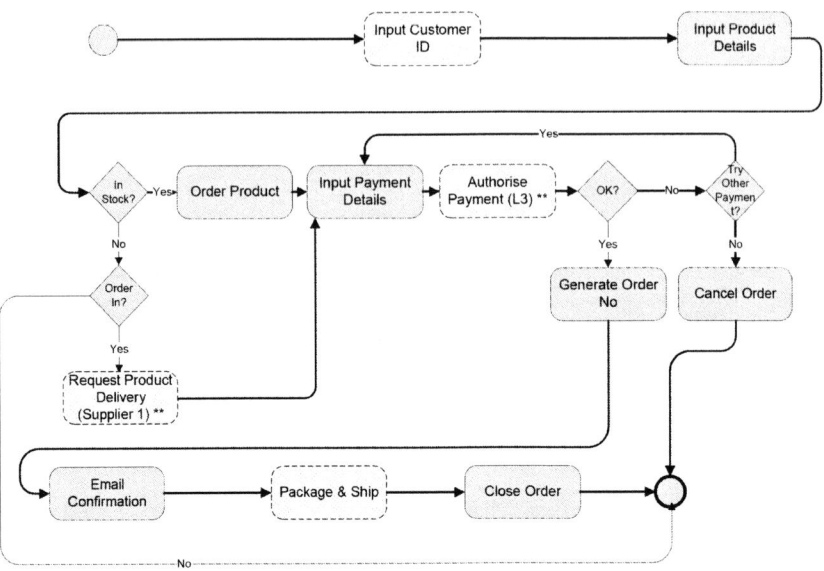

Fig. 6. Optimised Process — Flowchart

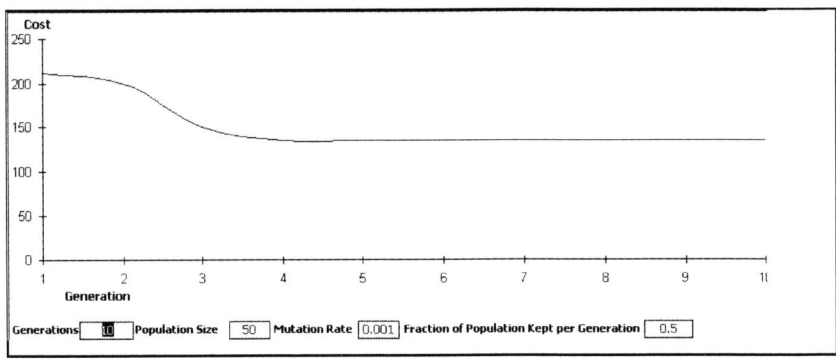

Fig. 7. Optimised Process — Cost per Generation Graph

and stored in the system, and the optimisation process is executed). Although the research behind this technique has been conducted in the service sector, there is no reason why it cannot be applied to any type of process (e.g. manufacturing processes).

6   Conclusion

The ideas presented in this paper are a result of the research question "how can a business process be represented as a chromosome?". Research into techniques used in group technology, cellular manufacturing and classification and coding systems led to a novel type of numerical process model. The purpose of this representation is to allow optimisation of a process via the use of GA. The actual optimisation technique, a hybrid combination of rule-based and GA programming, was also demonstrated in this paper and led to the optimisation of a telephone order process.

References

1. Johansson, H.J., McHugh, A.P., Pendlebury, J., Wheeler, W.A.: Business Process Reengineering Breakpoint Strategies for Market Dominance. Wiley, Chichester (1993)
2. Stelling, M.T., Roy, R., Tiwari, A., Majeed, B.: A Coding Mechanism for Business Process Optimisation Using Genetic Algorithms: Initial Thoughts. In: 10th IASTED International Conference on Artificial Intelligence and Soft Computing, Palma, Majorca. ACTA Press (2006)
3. Zakarian, A., Kusiak, A.: Analysis of Process Models. IEEE Transactions on Electronics Packaging Manufacturing 23(2), 137–147 (2000)
4. Gordijn, J., Akkermans, J.M., van Vliet, J.C.: What's in an Electronic Business Model. In: 12th International Conference on Knowledge Engineering and Knowledge Management EKAW 2000, Juan-les-Prins, October 2-6 (2000)
5. Object Management Group (OMG), Business Process Modelling Notation (BPMN) Specification: Final Adopted Specification, OMG, Needham, MA, USA (2006)
6. Vergidis, K.: Multi-Objective Optimisation of Business Processes Using Soft Computing Techniques (MSc Thesis), Cranfield University (2005)
7. Tiwari, A.: Evolutionary Computing Techniques for Handling Variables Interaction in Engineering Design Optimisation (PhD Thesis), Cranfield University (2001)
8. Zakarian, A.: Analysis of Process Models: A Fuzzy Logic Approach. International Journal of Advanced Manufacturing Technology (17), 444–452 (2001)
9. Phalp, K., Shepperd, M.: Quantitative Analysis of Static Models of Processes. The Journal of Systems and Software 52, 105–112 (2000)
10. Hofacker, I., Vetschera, R.: Algorithmical Approaches to Business Process Design. Computers & Operations Research 28, 1253–1275 (2001)
11. National Archives of Australia, Business Classification Systems (2000) (accessed 2006), http://www.naa.gov.au/recordkeeping/control/tools/appendixB.html
12. Oxford English Dictionary (accessed 2006), http://www.askoxford.com
13. Wiktionary (accessed 2006), http://en.wiktionary.org

Use of Multi-objective Evolutionary Algorithms in Extrusion Scale-Up

José António Covas and António Gaspar-Cunha

IPC- Institute for Polymers and Composites / I3N, University of Minho,
Campus de Azurém 4800-058 Guimarães, Portugal
{jcovas,agc}@dep.uminho.pt

Abstract. Extrusion scale-up consists in ensuring identical thermo-mechanical environments in machines of different dimensions, but processing the same material. Given a reference extruder with a certain geometry and operating point, the aim is to define the geometry and operating conditions of a target extruder (of a different magnitude), in order to subject the material being processed to the same flow and heat transfer conditions, thus yielding products with the same characteristics. Scale-up is widely used in industry and academia, for example to extrapolate the results obtained from studies performed in laboratorial machines to the production plant. Since existing scale-up rules are very crude, as they consider a single performance measure and produce unsatisfactory results, this work approaches scale-up as a multi-criteria optimization problem, which seeks to define the geometry/operating conditions of the target extruder that minimize the differences between the values of the criteria for the reference and target extruders. Some case studies are discussed in order to validate the concept.

Keywords: Multi-Objective Evolutionary Algorithms, Extrusion, Scale-Up.

1 Introduction

Scale-up is very often the action of defining the geometry and operating conditions of a machine that reproduce the working conditions of another of the same type and of different size, but processing the same material. This is a procedure of great practical importance. For example, in the case of polymer extrusion, scale-up rules are used to design large extruders using the results of studies performed on laboratory-scale machines. Extrapolating know-how instead of performing research on large-output machines allows for significant time savings [1-3].

Scale-up rules were proposed over several decades by different researchers, namely Carley and McKelvey (1953), Maddock (1959), Pearson (1976), Yi and Fenner (1976), Schenkel (1978), Chung (1984) and Rauwendaal (1986) [1-3]. These studies used analytical process descriptions to correlate large and small primary scaling variables (diameter, channel depth, screw length and screw speed) simply in terms of an exponent of their diameter ratio. However, since plasticating extrusion is a complex process involving solids conveying, melting of these solids and melt conveying, as well as other related phenomena such as mixing, such correlations only hold when a single process criterion is kept constant, e.g. constant melt flow shear rate, or constant melting rate. This was recognized in the reviews prepared by Rauwendaal (1987) and

E. Avineri et al. (Eds.): Applications of Soft Computing, ASC 52, pp. 86–94.

Potente (1991) [2, 3], who anticipated unbalanced solids and melt conveying rates when applying most of the rules available.

A more performing scaling-up methodology is therefore needed. It is important to consider simultaneously several process criteria and, since they are often conflicting, to know the degree of satisfaction eventually attained. Flexibility in terms of the criteria selected is also important, in contrast with the available methods that provide relations only for specific performance measures. Thus, it makes sense to consider extrusion scale-up as a multi-objective optimization problem, where the purpose is to define the geometry/operating conditions of the target extruder that minimize the differences between the values of the criteria for the reference and target extruders. This work applies a Multi-Objective Evolutionary Algorithm (MOEA) methodology, previously developed by the authors, to perform that task. In fact, rather than selecting the best optimization method, the aim here is to characterize the problem, propose a methodology and ascertain the level of success of the solutions proposed.

The text is organized as follows. In section 2 we present the optimization methodology (which uses a MOEA) and the process modelling routine, both developed by the authors. Section 3 discusses one scale-up example, which is presented and solved. Finally, section 4 proposes some conclusions.

2   Multi-objective Scale-Up

2.1 Optimization Methodology

As stated above, extrusion scale-up consists in extrapolating the behaviour of a reference extruder to another of the same type, but of different size (denoted as target extruder). Thus, we know the geometry and processing conditions of the reference extruder and wish to define either the operating conditions (if the machine exists), or the geometry and operating conditions (if it is to be built/purchased) of the target extruder, in such way that the major performance measures of both machines are as similar as possible. This is seen here as an optimization problem where we seek to determine the geometry/operating conditions of the target extruder that minimize the differences in performance in relation to the reference extruder.

The multi-objective scale-up optimization methodology proposed includes the following steps:

1- Use the process flow modelling routine to predict the responses of the reference extruder under a specific set of operating conditions and polymer system;
2- Analyse the results and define the most important parameters to be adopted for scale-up;
3- Gather information on target extruder (geometry: screw external diameter and length/diameter ratio; operating range: screw speed, set temperatures);
4- Perform scale-up via minimization of the differences in performance between the two extruders (optimization criteria).

The method requires three basic routines: a modelling package, a multi-objective optimization algorithm and a criteria quantification routine (see Figure 1). The algorithm defines automatically the (increasingly more performing) solutions to be used by the modelling routine. The parameter values obtained from the latter serve as input data to the criteria quantification routine, which compares them with the equivalent ones for the reference extruder. This information is supplied to the optimization routine, which defines new improved solutions to be evaluated, the process being repeated until a stop criterion is reached.

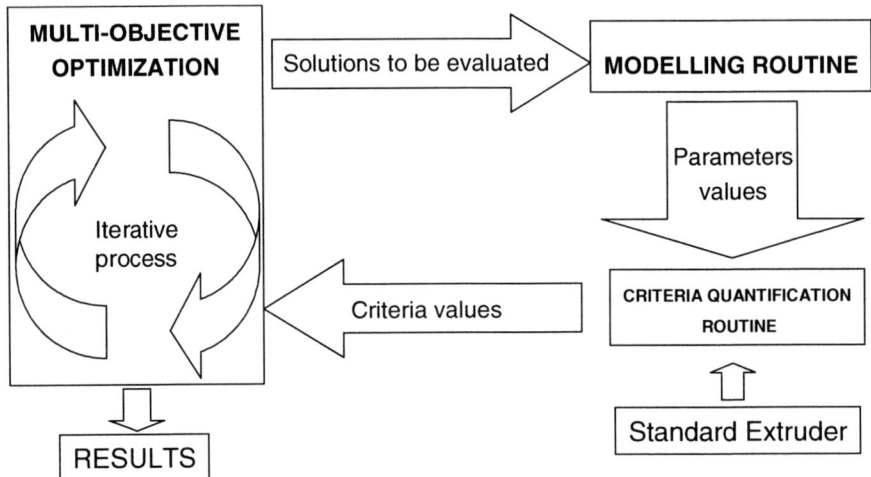

Fig. 1. Scale-up optimization methodology

2.2 Multi-objective Evolutionary Algorithms

During the last decade Multi-Objective Evolutionary Algorithms (MOEA) have been recognized as a powerful tool to explore and find out approximations to Pareto-optimal fronts in optimization problems [4, 5]. This is essentially due to their capacity to explore and combine various solutions to find the Pareto front in a single run and the evident difficulty of the traditional exact methods to solve this type of problems.

In a multi-objective algorithm the solution space is seen as sets of dominated and non-dominated points. These are solutions at least as good as the remaining with respect to all objectives, but strictly better with respect to at least one objective, *i.e.*, one solution point dominates another when it is equally good in every objective and formally better in at least one objective [4]. Since in MOEA the various criteria (or objectives) are optimized simultaneously, each individual solution belonging to the Pareto set establishes a compromise between all criteria. An efficient MOEA must distribute homogeneously the population along the Pareto frontier and improve the solutions along successive generations.

In this work, the Reduced Pareto Set Genetic Algorithm with elitism (RPSGAe) is adopted [6,7]. Initially, RPSGAe sorts the population individuals in a number of pre-defined ranks using a clustering technique, in order to decrease the number of solutions on the efficient frontier, while maintaining its characteristics intact. Then, the individuals' fitness is calculated through a ranking function. To incorporate this technique, the algorithm follows the steps of a traditional GA, except that it takes on an external (elitist) population and a specific fitness evaluation. Initially, the internal population is randomly defined and an empty external population is formed. At each generation, a fixed number of the best individuals, obtained by reducing the internal population with the clustering algorithm [6], is copied to the external population. This process is repeated until the number of individuals of the external population is complete. Then, the clustering technique is applied to sort the individuals of the external population, and a pre-defined number of the best ones is incorporated in the internal population, replacing the less fit individuals. Detailed information on this algorithm can be found elsewhere [6, 7].

2.3 Single Screw Extrusion Modelling Routine

Extrusion is a process whereby a molten polymer is forced to flow continuously through a die of a given shape, thus yielding a product with a constant cross-section (Figure 2). Despite the apparent simplicity of both machine and procedure, some basic functions must be accomplished if the product is to exhibit good performance. Process continuity is ensured by using an Archimedes-type screw, rotating inside the heated barrel at constant speed. Some screw geometric features and proper selection of barrel temperatures determine the most appropriate sequence of solid polymer conveying in the initial screw turns, progressive melting of this material, melt conveying with

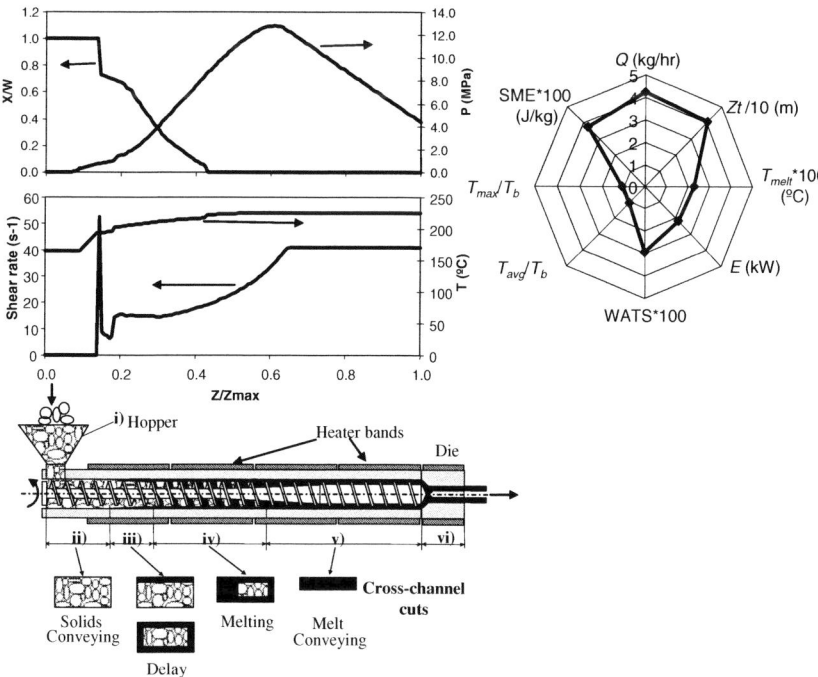

Fig. 2. Single screw extrusion: the machine, physical models and results of the modelling routine

pressure generation and flow through the die [1, 7]. These individual stages are also illustrated in Figure 2. Their characteristics are described mathematically by a set of differential flow equations, which are coupled by appropriate boundary conditions to provide a global plasticating model that is solved numerically.

The vertical pressure profile in the hopper is computed to set an initial condition at the extruder entrance. In the initial screw turns, we assume the linear displacement of an elastic solid plug subjected to increasing temperature due to the combined contribution of friction dissipation and heat conduction from the surrounding metallic surfaces. Delay (i.e., beginning of melting) is sub-divided into the initial existence of a melt film separating the solids from the barrel, followed by encapsulation of the solids by melt films. Melting follows a mechanism involving 5 distinct regions, one being the melt pool, another the solid plug and the remaining consist of melt films near to the channel walls. Melt pumping and die flow were modelled considering the non-isothermal flow of a non-Newtonian fluid. Calculations are performed in small screw channel increments, a detailed description being given elsewhere [7, 8].

2.4 Scale-Up Criteria

For scale-up purposes, it makes sense to define two types of criteria. The first deals with single value parameters such as power consumption (E), specific mechanical energy (energy consumption per unit output, SME), output (Q) or degree of mixing (weight average total strain, $WATS$), which are illustrated in the radar plot of Figure 2 and provide an overview of the extruder behaviour under a specific set of input conditions. Within the same type, other criteria could be selected, such as well-known adimensional numbers like Cameron, Peclet or Brinkman, which account for temperature development, relative importance of convection and conduction and extent of viscous dissipation, respectively, thus estimating complementary aspects of the thermo-mechanical environment. The second type of criteria deals with the evolution of certain parameters along the screw, such as melting (solid bed, X/W), pressure (P), shear rate ($\dot{\gamma}$) and temperature (T) axial profiles. The following equations are used to define the objective functions for single values and profile parameters, respectively (see Figure 3):

$$F_j = \frac{\left|C_j - C_j^r\right|}{C_j^r} . \tag{1}$$

$$F_j = \frac{\sum_{k=1}^{K} \dfrac{\left|C_{j,k} - C_{j,k}^r\right|}{C_{j,k}^r}}{K} . \tag{2}$$

where F_j is the fitness of criterion j, C_j and C_j^r are the values of criterion j (single values) for the target and reference extruders, respectively, and $C_{j,k}$ and $C_{j,k}^r$ are the values of criterion j on location k (along the extruder) for the target and reference extruders, respectively.

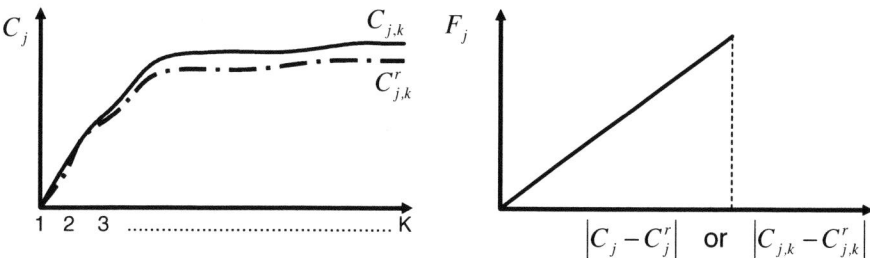

Fig. 3. Definition of the fitness of a criterion

3 Scaling-Up Single Screw Extruders

3.1 Example

Using as reference a laboratorial extruder with a diameter of 30 mm and as target an extruder with a diameter of 75 mm (see table 1 and Figure 4), we wish to perform scale-up in terms of operating conditions. Data for the reference extruder was obtained using a screw speed (N) of 50 rpm and a uniform barrel temperature (T_i) of 190 °C. The range of variation of the target extruder parameters is: N [10-200] rpm; T_i [170-230] °C. Data from polypropylene (NOVOLEN PHP 2150 from BASF) is adopted for the computational work.

Table 1. Geometry of the extruders used for scale-up

D (mm)	L/D	L_1/D	L_2/D	L_3/D	Compression ratio
75	30.0	10.0	10.0	10.0	3.3
30	30.0	10.0	10.0	10.0	2.5

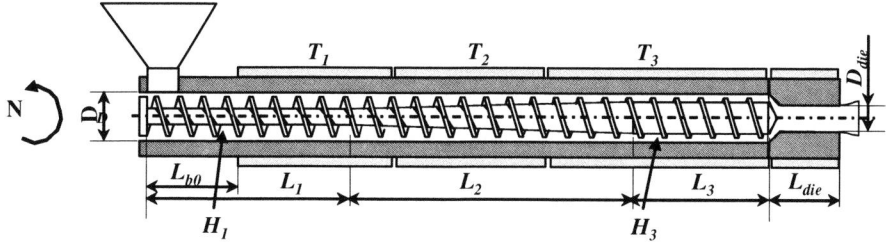

Fig. 4. Parameters required to describe the extruder geometry and operating conditions

3.2 Results and Discussion

Figure 5 shows the results of the optimization run when the various criteria were considered individually. A distinct set of operating conditions is proposed for each criterion. The smaller the value of the objective function, the more successful the scale-up is. As expected, scaling-up using criteria related to machine size (output, power consumption) becomes difficult for considerably different diameter ratios (in this case, 2.5). However, the use of constant values or functions related to flow characteristics (e.g., relative melting rate, average shear rate, average shear stress, viscous dissipation and adimensional numbers) is quite successful.

Figure 6 assesses the degree of satisfaction of the remaining criteria, when a specific single criterion is analysed. Shear rate, shear rate profile and Cameron number

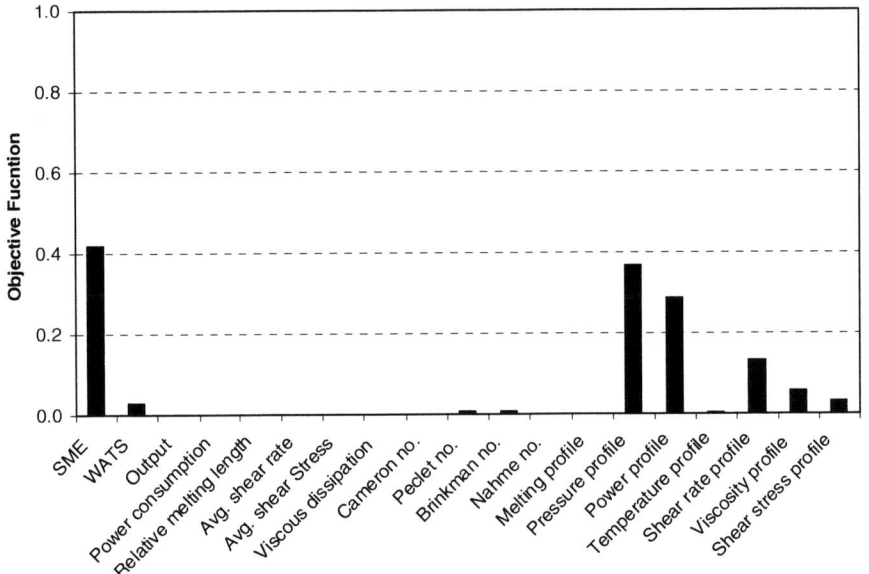

Fig. 5. Scaling-up for operating conditions using individual criteria

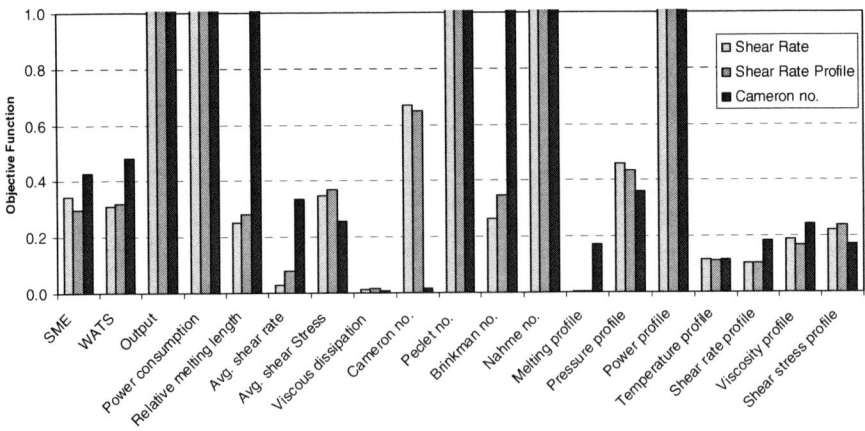

Fig. 6. Influence of the optimal operating conditions for shear rate, shear rate profile and Cameron no. on the satisfaction of the remaining criteria

were selected for this purpose. Not surprisingly, optimization of a single criterion is feasible, but has little value in terms of satisfying simultaneously other important performance measures which, in many cases, are conflicting.

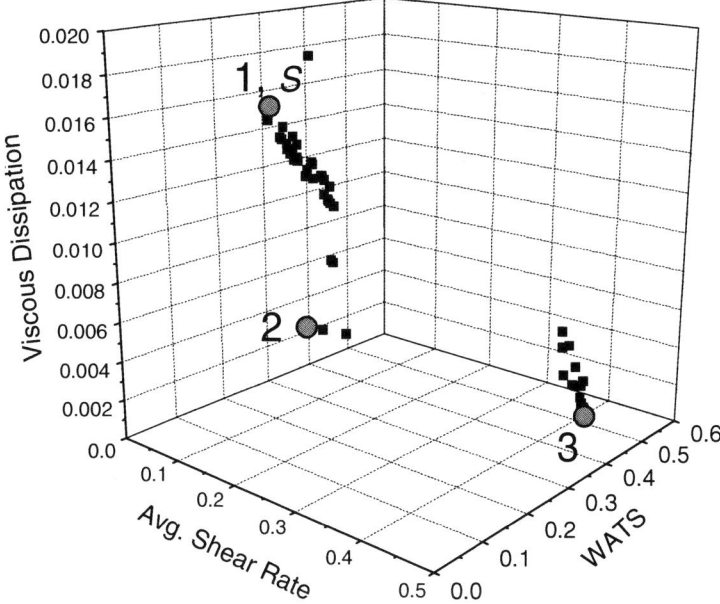

Fig. 7. Pareto frontiers for example 1

Table 2. Optimization with multiple criteria

Criteria	Point	N (rpm)	T_1 (°C)	T_2 (°C)	T_3 (°C)	F_1	F_2	F_3	F_4	F_5	F
C1,C2, C3	1	32.3	202.4	224.6	227.6	0.00	0.31	0.01	0.66	0.01	
	2	39.7	211.6	219.3	213.1	0.25	0.16	0.01	0.06	0.13	
	3	44.1	225.9	201.2	170.6	0.40	0.52	0.00	0.02	0.17	
	S	*32.3*	*202.4*	*224.6*	*227.6*	*0.00*	*0.31*	*0.01*	*0.66*	*0.01*	*0.20*
C1, C4 C5	1	32.3	226.1	200.9	213.5	0.00	0.31	0.01	0.66	0.01	
	2	40.5	224.1	189.5	206.2	0.27	0.43	0.01	0.00	0.16	
	3	34.4	223.2	200.5	199.7	0.06	0.31	0.01	0.68	0.00	
	S	*39.3*	*174.8*	*198.9*	*184.6*	*0.22*	*0.46*	*0.01*	*0.08*	*0.13*	*0.18*
C1 to C5	1	32.3	202.4	220.7	186.4	0.00	0.31	0.01	0.66	0.01	
	2	36.9	190.6	216.3	196.4	0.15	0.10	0.01	0.43	0.15	
	3	44.1	189.8	220.4	181.5	0.40	0.52	0.00	0.02	0.17	
	4	41.8	198.2	194.0	189.0	0.32	0.46	0.01	0.00	0.17	
	5	34.4	184.2	198.7	198.7	0.07	0.31	0.01	0.68	0.00	
	S	*36.9*	*190.6*	*216.3*	*196.4*	*0.15*	*0.10*	*0.01*	*0.43*	*0.15*	*0.17*

The advantages of multi-criteria optimization were tested with three examples. The first considers three criteria, average shear rate, C1, WATS (distributive mixing), C2, and viscous dissipation, C3. The second example deals with the simultaneous optimization of C1, Cameron number, C4, and melting profile, C5. Finally, the third example includes all criteria C1 to C5.

The results are shown in Figure 7 and Table 2. The figure presents the 3-dimensional Pareto surface for example 1, where criteria C1, C2 and C3 were optimized concurrently. The solutions identified as 1, 2 and 3 represent the best ones to minimize individually each criteria. Solution S minimizes the average of the 3 criteria, i.e., it yields a good compromise between the three criteria. Table 2 shows the operating conditions resulting from the solutions proposed for the three examples, the values of the 5 criteria (columns F1 to F5, but in examples 1 and 2 only three criteria were considered in the optimisation run) and the average F for solution S. When 5 criteria are optimised simultaneously a better solution is found, despite of the conflicting nature of some of the extruder responses.

4 Conclusions

The methodology proposed for extrusion scale-up is able to consider simultaneously various criteria and to take into account their relative importance. It can be applied to the scale-up of either operating parameters and/or geometry. Moreover, the efficiency of the scaling-up can be easily assessed by monitoring the implication of the exercise on the satisfaction of other process measures.

This methodology can be easily extended to other polymer processing technologies, as long as sufficiently precise modelling routines are available.

References

1. Rauwendaal, C.: Polymer Extrusion. Hanser Publishers, Munich (1986)
2. Rauwendall, C.: Scale-up of Single Screw Extruders. Polym. Eng. Sci. 27 (1987)
3. Potent, H.: Existing Scale-up rules for Single-Screw Plasticating extruders. Intern. Polym., Process. 6 (1991)
4. Deb, K.: Multi-Objective Optimization using Evolutionary Algorithms. Wiley, Chichester (2001)
5. Coello Coello, C.A., Van Veldhuizen, D.A., Lamont, G.B.: Evolutionary Algorithms for Solving Multi-Objective Problems. Kluwer, Dordrecht (2002)
6. Gaspar-Cunha, A., Covas, J.A.: RPSGAe - A Multiobjective Genetic Algorithm with Elitism: Application to Polymer Extrusion. In: Gandibleux, X., Sevaux, M., Sörensen, K., T'kindt, V. (eds.) Metaheuristics for Multiobjective Optimisation. Lecture Notes in Economics and Mathematical Systems. Springer, Heidelberg (2004)
7. Gaspar-Cunha, A.: Modelling and Optimization of Single Screw Extrusion, PhD Thesis, University of Minho, Guimarães, Portugal (2000)
8. Gaspar-Cunha, A., Covas, J.A.: The Design of Extrusion Screws: An Optimisation Approach. International Polymer Processing 16, 229–240 (2001)

## Part III

**Signal Processing and
Pattern Recognition**

A Multi-Agent Classifier System Based on the Trust-Negotiation-Communication Model

Anas Quteishat[1], Chee Peng Lim[1], Jeffrey Tweedale[2], and Lakhmi C. Jain[2]

[1] School of Electrical & Electronic Engineering
 University of Science Malaysia, Malaysia
[2] School of Electrical & Information Engineering
 University of South Australia, Australia

Abstract. In this paper, we propose a Multi-Agent Classifier (MAC) system based on the Trust-Negotiation-Communication (TNC) model. A novel trust measurement method, based on the recognition and rejection rates, is proposed. Two agent teams, each consists of three neural network (NN) agents, are formed. The first is the Fuzzy Min-Max (FMM) agent team and the second is the Fuzzy ARTMAP (FAM) agent team. An auctioning method is also used for the negotiation phase. The effectiveness of the proposed model and the bond (based on trust) is measured using two benchmark classification problems. The bootstrap method is applied to quantify the classification accuracy rates statistically. The results demonstrate that the proposed MAC system is able to improve the performances of individual agents as well as the team agents. The results also compare favorably with those from other methods published in the literature.

1 Introduction

The Multi-Agent System (MAS) approach has gained much research interest over the last decade. This is evidenced by the widespread application of MASs in different domains including eCommerce (Gwebu et al., 2005), healthcare (Hudson and Cohen, 2002), military support (Tolk, 2005), intelligent decision support (Ossowski et al., 2004), knowledge management (Singh et al., 2003), as well as control systems (Ossowski et al., 2002). A number of models have been used to describe the relation between agents in MASs, and one of the earliest models is the Beliefs, Desires, Intentions (BDI) reasoning model (Bratman, 1999). Another MAS model is the decision support pyramid model (Vahidov and Fazlollahi, 2004). The focus of this paper, however, is on the use of the Trust-Negotiation-Communication (TNC) model (Haider et al., 2006, Tweedale and Cutler, 2006) to implement MAS for pattern classification.

In our work, the proposed Multi-Agent Classifier (MAC) system consists of an ensemble of neural network (NN)-based classifiers. Two NN classifier agents are employed, (i) Fuzzy ARTMAP (FAM) (Carpenter et al., 1992) (ii) Fuzzy Min-Max (FMM) (Simpson, 1992). A novel method to measure trust by using the classification accuracy rates of the agent is proposed. To verify the effectiveness of the proposed MAC system and the trust measurement method, the Pima Indian Diabetes (PID) and the Wisconsin Breast Cancer (WBC) benchmark data sets are employed. The results

E. Avineri et al. (Eds.): Applications of Soft Computing, ASC 52, pp. 97–106.

are compared with those published in the literature. The bootstrap method is also applied to quantify the results statistically.

This paper is organized as follows. Section 2 gives a description on the architecture of the TNC-based MAC system. The proposed trust measurement and negotiation methods are explained in section 3. The experimental results and discussion are presented in section 4. Finally, section 5 gives a summary of the work presented in this paper.

2 A New TNC-Based Multi-Agent Classifier System

The TNC model used in this work is shown in Figure 1. The TNC model is based on the premise that the origin and the justification of the strength of beliefs come from the sources of belief. In this model, four possible sources of belief are considered: direct experience, categorization, reasoning, and reputation. In the TNC model, trust is used as a bond that can be strengthened via the exchange of certified tokens. In essence, trust is dynamic by nature; it increases by successful interactions, and is degraded by unsuccessful outcomes. Thus, measuring and quantifying trust is of prime importance to determine the success/failure of the TNC model.

Figure 2 shows the architecture of the proposed TNC-based MAC system. It consists of three layers. The top layer contains a parent agent who is responsible for making the final decision. The second layer contains team managers, while the third layer contains team members. As TNC is used as the framework of the proposed MAC system, methods for measuring trust and for making negotiation are necessary.

Negotiation used in the MAC system is based on an auctioning process. Auctioning is one of the most popular and mostly used negotiation processes in any MAS (Balogh et al., 2000). There are several bidding approaches, and the one used in this work is the "*sealed bid - first price auction*" method (Beer et al., 1999). In this negotiation method, each agent first submits its bidding value, without knowing the

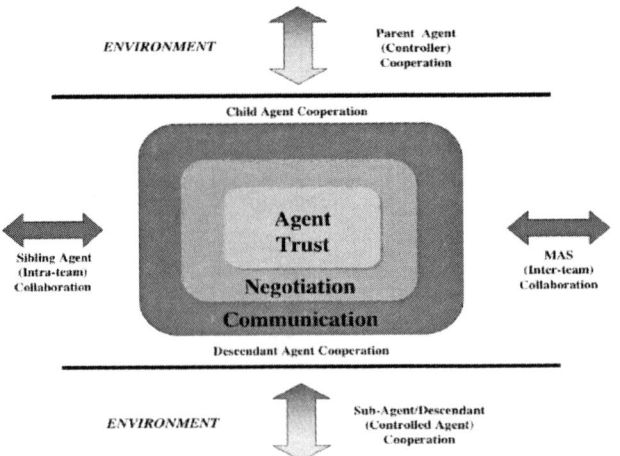

Fig. 1. The Trust-Negotiation-Communication (TNC) model

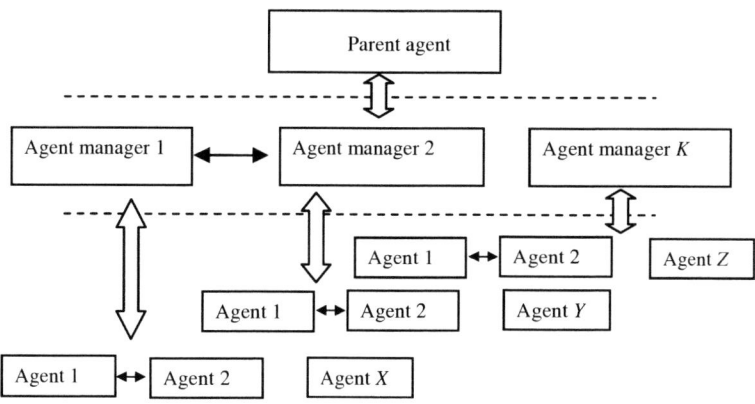

Fig. 2. The architecture of the proposed MAC system

bidding values of other agents (sealed bid). Then it is up to the auctioneer to take the decision based on the highest price.[1]

Given a new input sample, each agent within the team first gives a prediction (if available) of the output class of that input sample, and associates that prediction with a trust value. Then the team manager selects the prediction with the highest trust value. After that, an auctioning process takes place among the team managers. Each agent manager gives its prediction to the parent agent. Along with the prediction, each manager has to give a trust value, a reputation value, and a confidence factor. The parent agent makes a final decision and assigns a predicted output class for that particular input sample.

The proposed MAC system comprises two teams, each with three agents. The first team is formed by three FMM agents, while the second team is formed by three FAM agents. The FMM is a supervised classification network formed by using hyperbox fuzzy sets. A hyperbox defines a region of the n-dimensional pattern space that has patterns with full class membership. The hyperbox is completely defined by its minimum (min) and maximum (max) points. The membership function is defined with respect to these hyperbox min-max points, and describes the degree to which a pattern fits in the hyperbox. For an input pattern of n-dimensions a unit cube I^n is defined. In this case the membership value ranges between 0 and 1. A pattern which is contained in the hyperbox has the membership value of one.

On the other hand, the FAM network consists of a pair of Adaptive Resonance Theory (ART) modules designated as ART_a and ART_b, which create stable recognition categories in response to arbitrary sequences of input patterns. FAM also includes a map field module, F^{ab}, that establishes an association between input patterns and target classes. FAM uses fuzzy membership values between 0 and 1, indicating the extent to which each feature is present/absent, to represent its input patterns.

Although FAM and FMM have different network structures, they share a number of similarities. First, both networks possess the same incremental learning property,

---

[1] In future models, this would be modified to include other factors that contribute value to the bid. These factors have not been fully isolated as they can change with context.

i.e. they learn through a single pass of the training samples. Second, since both networks learn incrementally, the knowledge base formed in the networks is affected by the sequence of the training samples. In other words, different sequences of training samples form different knowledge bases in the network structures, hence different prediction capabilities. Third, in FMM, the weight vector of each node represents the minimum and maximum points that form a hyperbox in the pattern space. In FAM, the weight vector of each node also represents a hyperbox encoded by the minimum value of the input vectors and its complement-coded pairs. Finally, the performances of FMM and FAM primarily are governed mainly by one factor, i.e., the hyperbox size θ of FMM and the baseline vigilance parameter, $\overline{\rho_a}$ of FAM.

As mentioned earlier the performances of both NN models are affected by the orders/sequences of the training samples. One of the ways to improve the performances of these NN models is by voting (Carpenter et al., 1992). Instead of voting, we propose the use of agent teams to overcome this order/sequence problem. Here, each FMM/FAM agent is created by randomizing the sequence of the training samples. This satisfies the requirement that all agents within the team have the same learning capabilities, but with different initial knowledge bases.

3   Measuring Trust in the MAC System

As shown Figure 1, the core element of the TNC model is based on trust. There are many definitions for trust. A simple one is the confidence places in a person/thing, or more precisely the degree of belief in the strength, ability, truth, or reliability of a person/thing (Kelly et al., 2003). Measuring trust is inherently subjective; therefore a process of monitoring its attributes must be established. In this paper we propose to use the reliability rate (Xu et. al., 1992) for trust calculation. The reliability rate of each FMM/FAM agent is calculated using equation (1):

$$Trust = Reliability = \frac{Recognition}{1 - Rejection} \qquad (1)$$

where recognition is the ratio of the number of correct classifications to the total number of samples, and rejection is the ratio of the number of rejected classifications to the total number of samples.

In the original manifestation of FMM and FAM, there is no rejection criterion. To calculate the rejection rate, we propose to use the failure quality concepts discussed in Egmont-Petersen et al. (1994). Two suitable failure quantities used for the rejection criterion include interference and the restrictedness, and they are defined as:

- **Interference** being the degree to which a classifier is unable to label a class to an input sample because the sample seems to fit in more than one class;
- **Restrictedness** is the degree where a classifier leaves an input sample without classification because that sample does not fit to any class label.

To measure the above quantities, we apply the weighted confidence factor (CF) method, as proposed in Carpenter and Tan (1995), for both FAM/FMM agents. Based on the confidence factor, the failure concepts are therefore calculated using equations (2) and (3):

$$Interference = \begin{cases} 1 & if \; |CF_1 - CF_2| < \alpha \quad class(CF_1) \neq class(CF_2) \\ 0 & otherwise \end{cases} \quad (2)$$

$$Restricted\,ness = \begin{cases} 1 & if \; CF_1 < \beta \\ 0 & otherwise \end{cases} \quad (3)$$

where CF_1 and CF_2 are the confidence factors of the highest and second highest responded F_2^a node/hyperbox activated by the current input sample, and α is a user-defined interference threshold and β is a user-defined restrictedness threshold.

To calculate the confidence factors, the training data set is divided into two subsets: one for training and another for prediction, as proposed in Carpenter and Tan (1995). The prediction data set is then used to calculate the confidence factor for each F_2^a node/hyperbox. In addition, we use the prediction data set to calculate the initial trust for each agent. The initial trust is referred to as the reputation of an agent. The overall team reputation is the average of the reputations for all agents within the team.

To preserve the dynamic nature of trust measurement, the trust value is increased with successful predictions and is reduced with unsuccessful ones. To satisfy this nature of trust measurement, if each time the prediction of a sample is not rejected and is correctly classified, the trust value is increased by the *Reliability* value divided by the total number of test samples. However, if a sample is wrongly predicted or a sample is rejected for classification, the trust value is reduced by a factor equal to one over the total number of test samples.

In the test phase, the parent agent (the auctioneer) provides the current test sample to the team managers. The managers propagate the test sample to all team agents. Each agent gives a prediction for that particular sample to the team manager along with the trust value of that prediction and the confidence factor of the F_2^a node/hyperbox responsible for that prediction. The team manager then chooses the prediction with the highest trust value, and submits this prediction to the parent agent along with its reputation value, the trust value of the prediction, and the confidence factor for responsible F_2^a node/hyperbox. When the parent agent receives predictions from all teams, it makes a final prediction based on the highest Decision value using equation (4), as follows.

$$Decision = Team\;Reputation + Trust\;value + CF\;value \quad (4)$$

In the next section, an empirical evaluation using two benchmark data sets are presented in order to demonstrate the effectiveness of the proposed TNC-based MAC system and verify the proposed technique used to measure trust.

4 Experiments and Results

The proposed method is evaluated two benchmark[2] data sets. After several trial runs, the free parameter setting of the MAC system was as follows: the hyperbox size (θ) of

---

[2] The Pima Indian Diabetes (PID) and Wisconsin Breast Cancer (WBC) data sets from the UCI machine learning repository (Asuncion & Newman, 2007).

FMM was set to 0.275, while the baseline vigilance (ρ_a) of FAM was set at 0.70. The interference threshold α was set to 0.05 and the restrictedness threshold β was set to 0.1. The experiment was conducted ten times for agent team. During each run, the order/sequence of the training data samples for each FAM and FMM agent was randomized. The final TNC-based test accuracy rate was calculated using equation (5):

$$TNC\text{-}based\ test\ accuracy = \frac{Number\ of\ correctly\ classifid\ test\ samples}{Total\ number\ of\ test\ samples - (\ number\ of\ non\text{-}predicted\ test\ samples)} \quad (5)$$

4.1 The PID Problem

The PID data set consists of 768 cases which belong to two classes, in which 268 cases (35%) are from patients diagnosed as diabetic and the remaining are healthy. The experiment was conducted using 60% of the data set for training, 20% for prediction and the remaining 20% for test. To quantify the performance statistically, the bootstrap method was used to compute the mean accuracy across all 10 runs.

Table 1 shows the bootstrap mean results of the FMM and FAM teams. The performance of each agent under the TNC framework is also shown. Note that there were some non-predictions from each agent under the TNC framework. It can be seen that each agent under the original setting or the TNC framework provides comparable results. However, the overall team accuracy rate can be further improved based on the aggregated results.

Table 1. Boot strapped results of the FMM and FAM teams for the PID problem

	FAM NN			FMM NN		
	Original FAM	TNC-based system		Original FAM	TNC-based system	
	Accuracy	Accuracy	Non-predictions	Accuracy	Accuracy	Non-predictions
Agent 1	71.79	73.13	1-21	69.47	69.44	1-12
Agent 2	68.05	69.04	7-30	68.89	68.90	1-15
Agent 3	71.20	72.39	3-18	67.36	67.20	1-13

Figure 3 shows the accuracy rates of the FMM and FAM teams, as well as the overall MAC system. The error bars indicate the 95% confidence intervals of the results estimated using the bootstrap method. Note that the accuracy rate of FMM is 4.38 % higher than the highest accuracy rate from single FMM (Table 1). The FAM team accuracy rate is 3.39% lower than the maximum single agent accuracy. However, the overall MAC system accuracy rate is higher than those of the FMM and FAM teams. Another important observation is that the MAC system managed to classify all the input samples (i.e. no non-predictions). The FMM team classified an average of 70.708% of all test samples while the FAM team classified the remaining samples. The results show the benefit of the proposed MAC system, in which input samples that cannot be handled by one team can be classified by another team, hence improvement in the overall performance.

4.2 The WBC Problem

The WBC data set contains 699 records of virtually assessed nuclear features of fine needle aspirates from patients, with 458 (65.5%) benign and 241 (34.5%) malignant

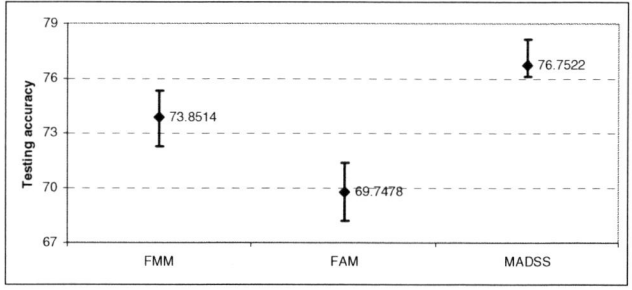

Fig. 3. Results of the FMM and FAM teams and the MAC system for PID

cases of breast cancer. Again, the training, prediction, and test sets, respectively, comprised 60%, 20%, and 20% of the total data samples. Like the previous case, the experiment was conducted 10 times. In each new run, the order/sequence of the training data samples for each FAM/FMM agent was randomized.

Table 2 shows the bootstrap results of the FMM and FAM teams. It can be seen that the results of each agent in the TNC-based system are better than those from single FMM agents, but inferior than those from single FAM agents. Notice that there are a lot of non-predictions from each FAM agent under the TNC framework. In other word, it is conservative in yielding predictions.

Table 2. Bootstrapped results of the FMM and FAM teams for the WBC problem

	FAM NN			FMM NN		
	Original FAM	TNC-based system		Original FAM	TNC-based system	
	Accuracy	Accuracy	Non-predictions	Accuracy	Accuracy	Non-predictions
Agent 1	95.71	92.31	0-89	95.14	97.69	13-31
Agent 2	95.71	95.56	3-48	93.57	96.77	13-33
Agent 3	95.00	87.80	0-24	92.14	97.52	6-29

Figure 4, shows the test accuracy rates of the FMM and FAM teams, as well as the MAC system. Again, the error bars indicate the 95% confidence interval of the results (estimated using the bootstrap technique). For the MAC system, out of the 140 test samples, the FMM team classified 104 samples while the FAM team classified the remaining 36 samples. Again, the results demonstrate the benefit of the proposed MAC system, i.e. the agent teams are able to cover the shortcomings of each other, and to produce an improved performance under the TNC framework.

4.3 Performance Comparison

To compare the effectiveness of the proposed MAC system, its results are compared with those published in the literature. For both PID and WBC data sets, a number of NN and machine learning techniques have been evaluated and reported in Hoang (1997). As shown in Table 3, the proposed MAC system is able to achieve the best test accuracy rates for both problems. This ascertains the effectiveness of the proposed MAC model and the associated trust measurement method. However, the

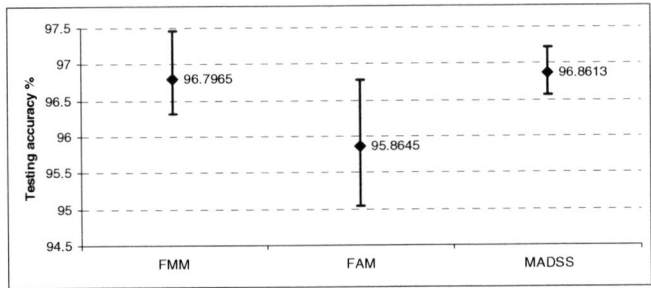

Fig. 4. Results of the FMM and FAM teams and the MAC system for WBC

shortcoming of the MAC system is that it requires 6 times the CPU time of single FMM/FAM network. Note that single FMM/FAM network used less than 10 sec on a Pentium Centrino 1.73 GHz computer for learning, based on the PID and WBC data sets.

Table 3. Performance Comparison of different methods for the PID and WBC problems. The results (except MAC) are extracted from Hoang (1997).

Method	PID Accuracy (%)	WBC Accuracy (%)
C4.5	71.02	94.25
C4.5 rules	71.55	94.68
ITI	73.16	91.14
LMDT	73.51	95.75
CN2	72.19	94.39
LVQ	71.28	94.82
OC1	50.00	93.24
Nevprop	68.52	95.05
MAC (our proposed system)	76.75	96.86

5 Summary

In this paper, a new MAC system, based on the TNC model, has been proposed. It has been shown that "trust" is the core of the TNC model. As such, a novel method for trust measurement has been introduced. The method is based on the recognition and rejection rates of classification. To investigate the effectiveness of the trust measurement methods, two NN-based agent teams, i.e., FMM and FAM, were formed. Two benchmark problems were used to evaluate the applicability of the proposed model. The results showed that the TNC-based MAC system yielded better test accuracy rates than those from individual teams, as well as those from a number of machine learning systems published in the literature.

   Although the results from the benchmark studies are encouraging, more experiments with data sets from different domains are needed to further ascertain the effectiveness of the proposed MAC system and the trust measurement method. Besides, instead of FMM/FAM, use of other classifiers for the proposed TNC-based model can also be investigated.

References

Asuncion, A., Newman, D.J.: UCI Machine Learning Repository. University of California, Department of Information and Computer Science, Irvine (2007),
http://www.ics.uci.edu/~mlearn/MLRepository.html

Balogh, Z., Laclavik, M., Hluchy, L.: Multi Agent System for Negotiation and Decision Support. In: Proceeding of fourth International Scientific Conference Electronic Computers and Informatics, Košice - Herľany, Slovakia, pp. 264–270 (2000)

Beer, M., D'inverno, M., Jennings, N., Luck, M., Preist, C., Schroeder, M.: Negotiation in Multi-Agent Systems. Knowledge Engineering Review 14, 285–289 (1999)

Bratman, M.E.: Intention, Plans, and Practical Reason. University of Chicago Press (1999)

Carpenter, G., Tan, A.: Rule extraction: From neural architecture to symbolic representation. Connection Science 7, 3–27 (1995)

Carpenter, G.A., Grossberg, S., Markuzon, N., Reynolds, J., Rosen, D.B.: Fuzzy ARTMAP: A neural network architecture for incremental learning of analog multidimensional maps. IEEE Trans. Neural Networks 3, 698–713 (1992)

Egmont-Petersen, M., Talmon, J.L., Brender, J., Ncnair, P.: On the quality of neural net classifiers. Artificial Intelligence in Medicine 6, 359–381 (1994)

Gwebu, K., Wang, J., Troutt, M.D.: Constructing a Multi-Agent System: An Architecture for a Virtual Marketplace. In: Phillips-Wren, G., Jain, L. (eds.) Intelligent Decision Support Systems in Agent-Mediated Environments. IOS Press, Amsterdam (2005)

Haider, K., Tweedale, J., Urlings, P., Jain, L.: Intelligent Decision Support System in Defense Maintenance Methodologies. In: International Conference on Emerging Technologies ICET 2006, pp. 560–567 (2006)

Hoang, A.: Supervised Classifier Performance on the UCI Data Set. Department of Computer Science. Australia, University of Adelaide (1997)

Hudson, D.L., Cohen, M.E.: Use of intelligent agents in the diagnosis of cardiac disorders. In: Computers in Cardiology, pp. 633–636 (2002)

Kelly, C., Boardman, M., Goillau, P., Jeannot, E.: Guidelines for trust in future ATM systems: A literature review. Technical Report 030317-01, European Organization for Safety of Air Navigation (2003)

Ossowski, S., Fernandez, A., Serrano, J.M., Hernandez, J.Z., Garcia-Serrano, A.M., Perez-De-La-Cruz, J.L., Belmonte, M.V., Maseda, J.M.: Designing multiagent decision support system the case of transportation management. In: Proceedings of the Third International Joint Conference on Autonomous Agents and Multiagent Systems AAMA, pp. 1470–1471 (2004)

Ossowski, S., Hernandez, J.Z., Iglesias, C.A., Ferndndez, A.: Engineering agent systems for decision support. In: Third International Workshop Engineering Societies in the Agents World ESAW 2002, Madrid, Spain, pp. 184–198 (2002)

Simpson, P.K.: Fuzzy Min-Max neural networks-Part 1: Classification. IEEE Transactions on Neural Networks 3, 776–786 (1992)

Singh, R., Salam, A., Lyer, L.: Using agents and XML for Knowledge representation and exchange: An intelligent distributed decision support architecture. In: Proceeding of the Ninth American Conference on Information Systems, pp. 1853–1864 (2003)

Tolk, A.: An Agent-Based Decision Support System Architecture for the Military Domain. In: Phillips-Wren, G., Jain, L. (eds.) Intelligent Decision Support Systems in Agent-Mediated Environments, ISO Press (2005)

Tweedale, J., Cutler, P.: Trust in Multi-Agent Systems. In: Proceeding of the 10th International Conference on Knowledge-Based Intelligent Information and Engineering Systems, Bournemouth UK, pp. 479–485. Springer, Heidelberg (2006)

Vahidov, R., Fazlollahi, B.: Pluralistic multi-agent decision support system: a framework and an empirical test. Information and Management 41, 883–898 (2004)

Xu, L., Krzyzak, A., Suen, C.Y.: Methods of combining multiple classifiers and their applications to handwriting recognition. IEEE Trans Systems, Man, and Cybernetics 22, 418–435 (1992)

Error Compensation Based Directional Interpolation Algorithm for Video Error Concealment

Liyong Ma[1], Yude Sun[2], and Naizhang Feng[3]

[1] School of Information Science and Engineering, Harbin Institute of Technology at Weihai, Weihai, 264209, P.R. China
maliyong@hit.edu.cn
[2] School of Information Science and Engineering, Harbin Institute of Technology at Weihai, Weihai, 264209, P.R. China
sun-yude@163.com
[3] School of Information Science and Engineering, Harbin Institute of Technology at Weihai, Weihai, 264209, P.R. China
fengnz@yeah.net

Summary. Spatial interpolation scheme is efficient to recovery lost blocks with abrupt scene changes or irregular motion, and directional interpolation algorithm has been widely used in video error concealment. An novel error compensation directional interpolation algorithm based on least squares support vector machines (LS-SVM) for video error concealment is proposed. LS-SVM is trained with the directional interpolation error distribution of neighbor block pixels around the missed block. Interpolation error compensation is employed to the interpolated result of missed pixels with LS-SVM estimation to obtain more accuracy estimation result. Experimental results demonstrate that the proposed algorithm is more efficient than the directional interpolation algorithm and linear algorithm.

Keywords: error concealment, support vector machines (SVM), error compensation, interpolation.

1 Introduction

In recent years there has been considerable interest in video error concealment. Multimedia information that is compressed to transport may be altered or lost due to channel noise. This kind of transmission errors leads to useless decoded information and visual artifacts that are not acceptable for certain applications. Except traditional error control and recovery techniques, error concealment schemes is developed to produce least objectionable output signal that is a close approximation to the original signal [11] . Many postprocessing error concealment schemes at the decoder have been proposed by employing the fact that human eyes are tolerate to certain distortion in high frequency. In there schemes interpolation is usually employed to recover damaged or lost blocks after error detection.

E. Avineri et al. (Eds.): Applications of Soft Computing, ASC 52, pp. 107–114.
springerlink.com   © Springer-Verlag Berlin Heidelberg 2009

The well-known approaches to image interpolation are nearest neighbor interpolation, linear interpolation and cubic interpolation [11] [8] [10]. However these methods blue images particularly in edge regions. Many other interpolation algorithms were proposed to deal with this problem. Among these interpolation algorithms, directional interpolation algorithm [3] has been widely used in video error concealment, for example, in [9], [4] and [1].

Most interpolation algorithms employ source images interpolation to establish result images without error compensation. However error compensation approaches are usually efficient to improve interpolation accuracy of result images. An efficient interpolation algorithm based on support vector regression (SVR) was introduced in [7]. Another improved error correction scheme was proposed in [5]. Source images are firstly down-sampled and then interpolated to get interpolation error with support vector machines training. After interpolation, result images are corrected with estimated error employing support vector regression. The experiments showed that the error correction approach was efficient for linear, cubic and other adaptive interpolation algorithms. A similar approach applying neural network is introduced in [6].

In this paper a novel error compensation directional interpolation algorithm based on least square support vector machines (LS-SVM) is proposed for video error concealment. Experimental results show that the result images of error compensation directional interpolation algorithm have higher quality than that of directional interpolation one. Background on LS-SVM is discussed in Section 2. In Section 3 the proposed error compensation algorithm is detailed after the introduction of directional interpolation algorithm. The experimental results are discussed in Section 4. Finally concluding remarks are provided in Section 5.

This work has been partially supported by Natural Science Foundation of Shandong Province for work on the learning based image interpolation study. This work was also partially supported by Study Fundation of Harbin Institute of Technology at Weihai under grant HIT(WH)200723.

2   Least Square Support Vector Machines

Support Vector Machines have been used successfully for many supervised classification tasks, regression tasks and novelty detection tasks [2]. A wide range of image processing problems have also been solved with SVMs as a machine learning tool.

The training set of SVMs in which each example is described by a d- dimensional vector, $x \epsilon \mathbb{R}^d$, consists of n training examples. The labels are used to describe categories that training examples belonging to. Following training, the result is an SVM that is able to classify previously unseen and unlabeled instances into a category based on examples learnt from the training set.

Support vector regression (SVR) is a function approximation approach applied with SVMs. A training data set consists of n points $\{x_i, y_i\}$, $i = 1, 2, .., n$, $x_i \epsilon \mathbb{R}^d$, $y_i \epsilon \mathbb{R}^d$, where x_i is the i-th input pattern and y_i is the i-th output pattern.

The aim of SVR is to find a function $f(x) = w \cdot x + b$, under the constrains $y_i - w \cdot x - b \le \varepsilon$ and $w \cdot x + b - y_i \le \varepsilon$ to allow for some deviation ε between the eventual targets y and the function $f(x)$ to model the data. By minimizing $\|w\|^2$ to penalize over-complexity and introducing the slack variables ξ_i, ξ_i^* for the two types of training errors, the regression weight results can be reached. For a linear ε-insensitive loss function this task therefore refers to minimize

$$\min \quad \|w\|^2 + C \sum_{i=1}^{n} \xi_i + \xi_i^*, \tag{1}$$

subject to $y_i - w \cdot x - b \le \varepsilon + \xi_i$ and $w \cdot x + b - y_i \le \varepsilon + \xi_i^*$, where all the slack variables are positive.

For linearly non-separable case, a mapping function $\varphi : \mathbb{R}^d \to \mathbb{R}^s$ can be found to map the current space into a higher dimensional one in which the data point is separable. The dot product in the mapped space is avoided by kernel function $\psi(x, y)$ that can be selected as linear kernel, polynomial kernel, radial basis function kernel or two layer neural kernel. More details about SVMs can be found in [2].

For LS-SVM approach, a special loss function of error ξ_i is employed, so the optimization problem can be given as

$$\min \quad J(w, \xi) = \frac{1}{2} w^T \cdot w + \gamma \frac{1}{2} \sum_{i=1}^{n} \xi_i^2. \tag{2}$$

Then Lagrange function can be defined and assume that α_i is the Lagrange const, kernel function $K(x(i), x_j)$ satisfied with Mercer condition. Ls-SVM model can be defined as

$$f(x) = \sum_{i=1}^{n} \alpha_i K(x, x_i) + b. \tag{3}$$

The calculation of the above model is faster than the traditional SVM model, and less resource is used in the calculation. In this paper LS-SVM is used for machine learning of interpolation error.

3   Error Compensation Based Directional Interpolation Algorithm

The spatial interpolation scheme and the motion compensation scheme are two important approaches in error concealment. The motion compensation scheme is not successful with abrupt scene changes or irregular motion where spatial interpolation scheme is efficient to recovery lost blocks for nearby pixel values are highly correlated.

Source image is often segmented into blocks, for example each block with size 16×16 pixel, and encoded into packet which contains one or more blocks for transmission. It is assumed that appropriate transmission and detection schemes are employed, and lost or error packets and blocks which need to be restored can be located exactly. Spatial interpolation can be employed to restore missed pixels with neighbor pixels value. Directional interpolation can restore more details in the missing blocks with keep the edge integrity in visual perception [9].

3.1 Directional Interpolation

Directional interpolation approach is detailed with directional classification and spatial interpolation as follows.

Directional Classification
Often surrounding pixels of missed blocks suggest more than one edge direction for interpolation, so directional classification is employed to decide which direction is the most likely edge orientation used for interpolation. Sobel operator is used for gradient estimation to select one proper direction among eight directions:

$$S_x = \begin{bmatrix} -1 & 0 & 1 \\ -2 & 0 & 2 \\ -1 & 0 & 1 \end{bmatrix}, S_y = \begin{bmatrix} 1 & 2 & 1 \\ 0 & 0 & 0 \\ -1 & -2 & -1 \end{bmatrix}. \tag{4}$$

Sobel operator is applied to every pixel in neighbor blocks of missed block. After gradient measure is calculated, the gradient value is rounded to the nearest $22.5°$ that is one of the eight direction classification equally spaced around $180°$. The missed block and neighbor blocks with eight directions are illustrated in Figure 1. Finally a selection scheme is employed to choose one direction among the eight ones. The direction counter is increased if a line drawn through the pixels of the neighbor blocks with corresponding direction passes through the missed block. After voting of all the pixels of neighbor blocks, the direction with top counters can be selected as interpolation direction.

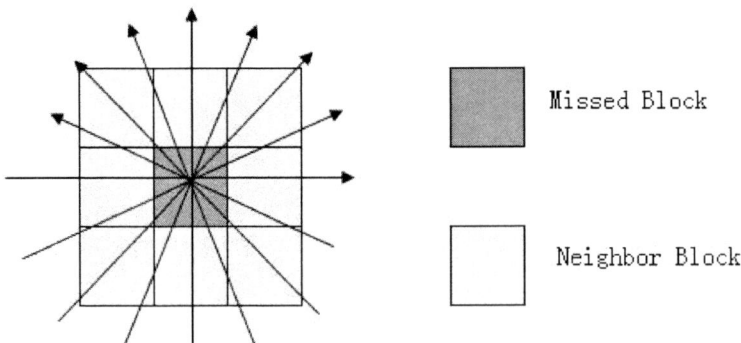

Fig. 1. Directional Interpolation

Spatial Interpolation

Every pixel value in the missed block is estimated with one-dimensional spatial interpolation of neighbor block pixels along the direction suggested by the classifier.

Let x_i be the i-th pixel in the missed block $X_{n \times n}$, and $y_j(x_i)$ the j-th pixel of the neighbor blocks of x_i in the corresponding line with suggested direction,where $i = 1, 2, ..., n \times n$, $j = 1, ..., m$. The estimated pixel value $\hat{x}_i$ of pixel x_i is calculated as

$$\hat{x}_i = \frac{\sum_{l=1}^{m} \frac{y_l(x_i)}{d_{il}^w}}{\sum_{k=1}^{m} \frac{1}{d_{ik}^w}}, \tag{5}$$

where d_{ik} is the distance from x_i to $y_k(x_i)$, w is the power of the inverse weighting and suggested value is 2.5 according to [9]. So this estimation value $\hat{x}_i$ can be used as the missed pixel value.

3.2   Proposed Interpolation Algorithm

Motivated by [6] and [5] where error compensation is very efficient for interpolation, another error compensation algorithm based on LS-SVM for directional interpolation is developed. Our proposed algorithm is described as follows.

(a) Directional classification is the same just as directional interpolation algorithm described in Section 3.1.

(b) Spatial interpolation

Every pixel value x_i of missed block $X_{n \times n}$ is calculated as follows:

(b1) Firstly estimated pixel value $\hat{x}_i$ is calculated with (5) just as directional interpolation.

(b2) One-dimension interpolation is performed to every pixel $y_j(x_i)$ with the other neighbor pixels that are used to estimate x_i except for the pixel $y_j(x_i)$. The interpolation is similar to the estimation of x_i in order to get the estimation of known pixels in the suggested interpolation direction of x_i. The estimation interpolation of $y_j(x_i)$ is calculated as

$$\hat{y}_j(x_i) = \frac{\sum_{l \neq j} \frac{y_l(x_i)}{d_{jl}^w}}{\sum_{k \neq j} \frac{1}{d_{jk}^w}}, \tag{6}$$

where d_{jk} is the distance from $y_j(x_i)$ to $y_k(x_i)$, w is the same power as in (5).

(b3) Interpolation error $\bar{y}_j(x_i)$ of every pixel $y_j(x_i)$ is calculated with

$$\bar{y}_j(x_i) = y_j(x_i) - \hat{y}_j(x_i), \tag{7}$$

where $y_j(x_i)$ is the known true value and $\hat{y}_j(x_i)$ the interpolated estimation value. Then the error distribution can be obtained with LS-SVM training. The input pattern of training set includes relative one-dimension coordinate d_{j1} and error value $\bar{y}_j(x_i)$ of every pixel. When the training finished, The trained LS-SVM could be employed to estimate interpolation error of x_i.

(b4) For every pixel x_i, interpolation error $\bar{x}_i$ is estimated with employing LS-SVM that has finished training in (b3). Input pattern for LS-SVM is the relative one-dimension coordinate, i.e. the relative distance between x_i and $y_1(x_i)$, and output is the estimated interpolation error. Then the final pixel value x_i can be obtained with the sum of $\hat{x}_i$ and $\bar{x}_i$, this error compensation value can be employed as missed pixel vale..

4   Experimental Results

We obtain similar results when the two directional interpolation algorithms and linear interpolation algorithm are employed to some standard video images, such as Foreman, Tennis and Flower Garden. The peak signal to noise ratio (PSNR) is compared with result images of different interpolation algorithms. PSNR for 8-bit gray image is defined as:

$$MSE = \frac{1}{MN} \sum_{m=0}^{M-1} \sum_{n=0}^{N-1} |\tilde{x}(m,n) - x(m,n)|^2, \tag{8}$$

$$PSNR = 10log\frac{255^2}{MSE}, \tag{9}$$

where the image size is $M \times N$, $\tilde{x}$ is the interpolation result image of x.

Results with linear interpolation, directional interpolation and error compensation based directional interpolation are employed to the test video images Foreman are displayed in Figure 2. And the detail regions of the result images are illustrated in Figure 3. It can be clearly found that the result images of the proposed interpolation approach is the best one among three approaches. The detail regions of the edge of the wall are kept closest to the source image. Different frames are compared in Table 1 and Table 2 with these error concealment approaches from video Foreman and Tennis. All the images are encoded and decoded with H.249 protocol. The proposed approach obtain the best PSNR value. It is shown that proposed error compensation algorithm improves the quality of the result images.

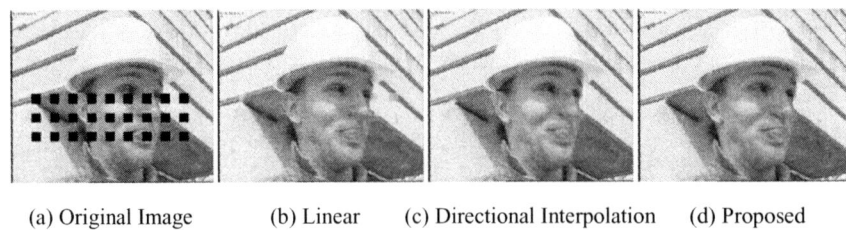

(a) Original Image   (b) Linear   (c) Directional Interpolation   (d) Proposed

Fig. 2. Interpolation Results with Different Approaches

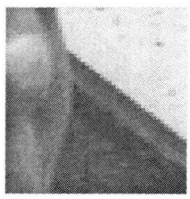 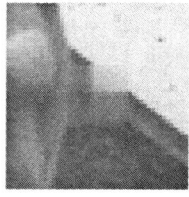 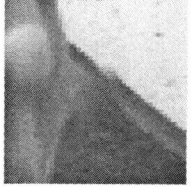 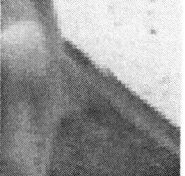

(a) Original Image   (b) Linear   (c) Directional Interpolation (d) Proposed

Fig. 3. Interpolation Local Results with Different Approaches

Table 1. PSNR of Foreman video error concealment

Frame no	Linear interpolation	Directional interpolation	Proposed interpolation
Frame 1	37.05	37.28	37.71
Frame 2	37.11	37.65	38.12
Frame 3	36.85	37.25	38.10
Frame 4	37.18	37.49	38.37
Frame 5	37.60	37.83	38.75
Frame 6	36.51	37.80	38.60
Frame 7	35.73	36.59	37.20
Frame 8	35.59	36.59	37.20
Frame 9	36.08	36.16	36.51
Frame 10	35.94	35.95	37.70
Average	36.56	37.06	37.70

Table 2. PSNR of Tennis video error concealment

Frame no	Linear interpolation	Directional interpolation	Proposed interpolation
Frame 1	30.79	31.45	32.01
Frame 2	30.81	31.51	32.17
Frame 3	30.88	31.76	32.44
Frame 4	30.87	31.51	32.02
Frame 5	30.88	31.65	32.22
Frame 6	30.88	31.57	32.08
Frame 7	30.89	31.76	32.49
Frame 8	30.89	31.67	32.24
Frame 9	30.89	31.68	32.52
Frame 10	30.86	31.64	32.24
Average	30.86	31.64	32.24

5 Conclusions

An error compensation directional interpolation algorithm based on LS-SVM has been proposed. The effectiveness of the algorithm for error concealment is confirmed by the experiments.

References

1. Agrafiotis, D., Bull, D., Canagarajah, N.: Spatial error concealment with edge related perceptual considerations. Signal Processing: Image Communication 21, 130–142 (2006)
2. Cristianini, N., Shawe-Taylor, J.: Introduction to support vector machines. Cambridge University Press, Cambridge (2000)
3. Kwok, W., Sun, H.: Multi-directional interpolation for spatial error concealment. IEEE Transactions on Consumer Electronics 39, 455–460 (1993)
4. Li, X., Orchard, M.: Novel sequential error-concealment techniques using orientation adaptive interpolaton. IEEE Transactions on Circuits and Systems for Video Technology 12, 857–864 (2000)
5. Liyong, M., Jiachen, M., Yi, S.: Support vector machines based image interpolation correction scheme. In: Wang, G.-Y., Peters, J.F., Skowron, A., Yao, Y. (eds.) RSKT 2006. LNCS, vol. 4062, pp. 679–684. Springer, Heidelberg (2006)
6. Liyong, M., Yi, S., Jiachen, M.: Neural network based correction scheme for image interpolation. In: Liu, D., Fei, S., Hou, Z., Zhang, H., Sun, C. (eds.) ISNN 2007. LNCS, vol. 4493, pp. 840–845. Springer, Heidelberg (2007)
7. Liyong, M., Yi, S., Jiachen, M.: Local spatial property based image interpolation scheme using SVMs. Journal of Systems Engineering and Electronics (in press)
8. Russ, J.: The image processing handbook, 4th edn. CRC Press, Boca Raton (2002)
9. Suh, J.W., Ho, Y.S.: Error concealment based on directional interpolation. IEEE Transactions on Consumer Electronics 43, 295–302 (1997)
10. Thevenaz, P., Blu, T., Unser, M.: Interpolation revisited. IEEE Transaction on Medical Imaging 19, 739–758 (2000)
11. Wang, Y., Zhu, Q.F.: Error control and concealment for video communication: A review. Proc. IEEE 86, 974–997 (1998)

An Evolutionary Approach to Digitalized Hand Signs Recognition

Ivanoe De Falco[1], Antonio Della Cioppa[2], Domenico Maisto[1],
Umberto Scafuri[1], and Ernesto Tarantino[1]

[1] Institute of High Performance Computing and Networking,
National Research Council of Italy
Via P. Castellino 111, 80131 Naples, Italy
{ivanoe.defalco,domenico.maisto}@na.icar.cnr.it,
{umberto.scafuri,ernesto.tarantino}@na.icar.cnr.it
[2] Natural Computation Lab, DIIIE, University of Salerno,
Via Ponte don Melillo 1, 84084 Fisciano (SA), Italy
adellacioppa@unisa.it

Abstract. The issue of automatically recognizing digitalized human–made hand signs is a crucial step in facing human–computer interaction and is of paramount importance in fields such as domotics. In this paper Differential Evolution is used to perform classification of hand signs collected in a reduced version of the Auslan database. The performance of the resulting best individual is computed in terms of error rate on the testing set and is compared against those of other ten classification techniques well known in literature. Results show the effectiveness and the efficacy of the approach in solving the recognition task.

Keywords: Differential evolution, classification, centroids, human–made hand signs, virtual reality, domotics.

1 Introduction

One of the most interesting tasks in artificial intelligence and machine learning is to let machines recognize human–made hand signs. Thanks to this, it will be possible to require the execution of tasks without the interaction by the classical I/O devices. This may have important influences in some high–technology application areas as for instance virtual reality, entertainment, domotics, real–time animation, etc. To fulfil this task a first step is to transform the records of human–made hand signs, captured by cameras, sensors or haptic gloves, into digitalized information which could be easily dealt with by a computer.

The second step is to classify these digitalized records. A classification tool [1, 2] is usually a part of a more general automatic pattern recognition system and aims at assigning class labels to the observations previously gathered by some sensor. To fulfil its task, a classifier relies on some features extracted in numeric or symbolic form by a feature extraction mechanism. The *supervised* classification scheme is based on the availability of a set of patterns that have

E. Avineri et al. (Eds.): Applications of Soft Computing, ASC 52, pp. 115–124.
springerlink.com   © Springer-Verlag Berlin Heidelberg 2009

already been classified or described (training set). In such a situation, starting from these presented data, the system has to guess a relation between the input patterns and the class labels and to generalize it to unseen patterns (testing set).

Since classification is a far from trivial task, several heuristic methods have been designed. In this paper Differential Evolution (DE) [3], a version of an Evolutionary Algorithm [4], is considered. The aim is to evaluate DE effectiveness and efficiency in performing a centroid–based supervised classification by taking into account a database composed by human–made hand signs, digitally recorded and collected in a 13–class data set extracted from Auslan (Australian sign language) [5] database. In the following, the term centroid means simply "class representative" and not necessarily "average point" of a cluster in the multidimensional space defined by the number of features in the database items. The idea is to use DE to find the positions of the class centroids in the search space such that for any class the average distance of instances belonging to that class from the relative class centroid is minimized. Error percentage for classification on testing set is computed on the resulting best individual. Moreover, the results are compared against those achieved by ten well–known classification techniques.

Paper structure is as follows: Section 2 describes the Auslan database, Section 3 outlines DE scheme, while Section 4 illustrates the application of our evolutionary system to the classification problem. Section 5 reports on the actions needed to transform the original Auslan database into another which could be directly dealt with by the DE tool, the related results and the comparison against ten typical classification techniques. Finally Section 6 contains conclusions and future works.

2   Auslan Database

Auslan is the sign language of the Australian Deaf community and has about 4000 signs. However, there are many ways these signs can be modified to alter the meaning slightly and, even more powerfully, signs can be combined to give new signs. As regards the structure of signs in Auslan, there are several components: handshape, starting location, orientation and movement. By mixing and matching them, a wide variety of signs can be formed.

Quite recently, a set of 95 Auslan signs were captured [6] from a native signer using high–quality position trackers and instrumented gloves. The hand sign acquisition tool used is a two–hand system, and the information recorded for each hand are:

- x, y and z positions, in meters, relative to a zero point set below the chin.
- *roll* expressed as a value in [–0.5, 0.5] with 0 being palm down. Positive means the palm is rolled clockwise from the perspective of the signer.
- *pitch* expressed as a value in [–0.5, 0.5] with 0 being palm flat (horizontal). Positive means the palm is pointing up.

- *yaw* expressed as a a value in $[-1.0, 1.0]$ with 0 being palm straight ahead from the perspective of the signer. Positive means clockwise from the perspective above the signer.
- *thumb, forefinger, middle finger, ring finger* and *little finger* bend measures, each between 0 (totally flat) and 1 (totally bent).

Spatial and temporal resolutions are very high. Position and orientation are defined to 14–bit accuracy, yielding a positional error less than one centimeter and an angle error less than one half of a degree. Finger bend was measured with 8 bits per finger, of which probably 6 bits were usable once the glove was calibrated. The refresh rate of the complete system was close to 100 frames per second, and all signals had significantly low noise level.

To create the Auslan database [7], samples from a native Auslan signer were collected over a period of 9 weeks and stored in 9 subdirectories, each of which is captured in a different day and contains 3 samples of 95 different signs. There are 27 samples for each of the 95 signs, for a total of 2,565 instances. Each sign is stored in a single file which consists of as many lines as there are frames associated to the sign. Each frame consists of 22 whitespace–separated numbers representing the 22 parameters, captured by sensors, the first 11 being signals from the left hand and the latter 11 those from the right hand. Those signals last on average 57 frames, but they are quite different in length, as the representations of a same sign may be slower or faster, and as different signs may require different time spans: for instance the sign *no* is the fastest. Due to the above reasons, some items last about 50 frames and others about 65.

The only work in literature about the Auslan database can be found in [6]. The approach there used is based on temporal classification, namely each sign is considered as a stream composed by a set of channels. A *temporal learner* called *TClass* is designed and implemented, showing good results.

3   Differential Evolution

Differential Evolution is a stochastic, population–based optimization algorithm and uses vectors of real numbers as representations of solutions.

The seminal idea of DE is that of using vector differences for perturbing the genotype of the individuals in the population. Basically, DE generates new individuals by adding the weighted difference vector between two population members to a third member. If the resulting trial vector yields a better objective function value than a predetermined population member, the newly generated vector replaces the vector with which it was compared. By using components of existing population members to construct trial vectors, recombination efficiently shuffles information about successful combinations, enabling the search for an optimum to focus on the most promising area of solution space.

In more detail, given a minimization problem with m real parameters, DE faces it starting with a randomly initialized population $P(t = 0)$ consisting of n individuals each made up by m real values. In [3] the authors tried to come up with a sensible naming–convention, so they decided to name any DE strategy with a string like $DE/x/y/z$, where x is a string which denotes the vector to be perturbed (*best* = the best individual in current population, *rand* = a randomly chosen one, *rand–to-best* = a random one, but the current best participates in the perturbation too), y is the number of difference vectors taken for perturbation of x (either 1 or 2), while z is the crossover method (*exp* = exponential, *bin* = binomial). We have decided to perturb a random individual by using one difference vector and by applying binomial crossover, so our strategy can be referenced as $DE/rand/1/bin$. In it for the generic i–th individual in the current population three integer numbers r_1, r_2 and r_3 in $[1, n]$ differing one another and different from i are randomly generated. Furthermore, another integer number k in the range $[1, m]$ is randomly chosen. Then, starting from the i–th individual a new trial one i' is generated whose generic j–th component is given by:

$$x_{i',j} = x_{r_3,j} + F \cdot (x_{r_1,j} - x_{r_2,j}) \tag{1}$$

provided that either a randomly generated real number ρ in $[0.0, 1.0]$ is lower than a value CR (algorithm parameter in the same range as ρ) or the position j under account is exactly k. If neither is verified then a copy takes place: $x_{i',j} = x_{i,j}$. F is a real and constant factor in $[0.0, 1.0]$ which controls the magnitude of the differential variation $(x_{r_1,j} - x_{r_2,j})$, and is a parameter of the algorithm.

This new trial individual i' is compared against the i–th individual in current population and, if fitter, replaces it in the next population, otherwise the old one survives and is copied into the new population. This basic scheme is repeated for a maximum number of generations g.

4   DE Applied to Classification

Encoding. We have chosen to face the classification task by coupling DE with a centroids mechanism, hereinafter referred to as DE–C. Specifically, given a database with C classes and N attributes, DE–C should find the optimal positions of the C centroids in the N?-dimensional space, i.e. it should determine for any centroid its N coordinates, each of which can take on real values. With these premises, the i–th individual of the population is encoded as it follows:

$$(\mathbf{p}_i{}^1, \ldots, \mathbf{p}_i{}^C) \tag{2}$$

where the position of the j–th centroid $\mathbf{p}_i{}^j$ is constituted by N real numbers representing its N coordinates in the problem space:

$$\mathbf{p}_i{}^j = \{p_{1,i}^j, \ldots, p_{N,i}^j\} \tag{3}$$

Then, any individual in the population consists of $C \cdot N$ components, each of which is represented by a real value.

Fitness. Following supervised classification, also in our case a database is divided into two sets, a training one and a testing one. The automatic tool learns on the former and its performance is evaluated on the latter.

The fitness function ψ is computed as the sum on all the training set instances of the euclidean distance in N–dimensional space between the generic instance $\mathbf{x}_j$ and the centroid of the known class it belongs to according to database $(\mathbf{p}_i^{\mathbf{CL}_{known}(\mathbf{x}_j)})$. This sum is normalized with respect to D_{Train}, i.e., the number of instances which compose the training set. In formulae, the fitness of the i–th individual is given by:

$$\psi(i) = \frac{1}{D_{\text{Train}}} \cdot \sum_{j=1}^{D_{\text{Train}}} d\left(\mathbf{x}_j, \mathbf{p}_i^{\mathbf{CL}_{known}(\mathbf{x}_j)}\right) \qquad (4)$$

When computing distance, any of its components in the N–dimensional space is normalized with respect to the maximal range in the dimension, and the sum of distance components is divided by N. With this choice, any distance can range within $[0.0, 1.0]$ and so can ψ. Given the chosen fitness function, the problem becomes a typical minimization problem.

The rationale behind this fitness is that it can vary with continuity. In fact it varies even when centroid positions undergo small variations which might or not cause changes of class assignment for some instances of the training set.

Performance of a run, instead, is computed as the percentage $\%err$ of instances of testing set which are incorrectly classified by the best individual (in terms of the above fitness) achieved in the run. With this choice DE–C results can be directly compared to those provided by other classification techniques.

5   Experiments and Results

5.1   Preprocessing of Auslan Database

The Auslan database as it can be downloaded from [7] is not in a form which can be directly dealt with by any classification tool, yet it must be preprocessed.

Given the huge amount of data, the preliminary necessary action of collecting all these spread data into a single table is quite time consuming. With reference to this, it is important to note that not all of the 95 signs seem of interest for tasks related to domotics: in fact the high majority of the signs, like for instance *alive, building, boy, crazy, happy, juice, love, Norway, polite, sad, sorry, wild,* look of little or no interest for an application in domotics, thus they have not been taken into account. Rather, the attention has been focused on a subset of thirteen signs which may be important to let a human communicate with a computer, especially when this latter is supposed to carry out the task of performing actions useful to improve quality of life of the human user. This subset consists of the signs: *cold, drink, eat, exit, hello, hot, no, read, right, wait, write, wrong, yes,* which, therefore, constitute the thirteen classes of the reduced database for a total of $13 \cdot 27 = 351$ instances. The classes are numbered in such a way that the i–th class contains the instances of the i–th sign listed in the above shown alphabetical order, so that, for example, class 1 contains the items of the *cold* sign and class 5 those of *hello*.

To use the above described classification scheme, the idea would be to identify all the variables representing any database item by using all of the data of each frame, in their sequential order of appearance. Unfortunately, since the signs have different lengths, all the recordings must be transformed into ones having exactly the same number of frames. The intermediate choice has been made to cut all the recordings at exactly the average length of all signs, i.e. 57 frames. Hence, the longer ones have lost some of their last frames, whereas for those lasting less than 57 frames the hypothesis has been made that, since the sign is over, we can extrapolate it as if the hand stood still for the remaining frames. After doing so, the number of attributes for any database item would be $57 \cdot 22 = 1,254$, which is unmanageable for any classification tool.

A second choice has been taken to downsample the recordings by saving one frame every five. This is a quite common choice in the field and usually does not yield problems, provided that data are continuous as this is the case. By doing so, just 11 frames are considered for any sign.

A third choice has been to consider only the variables related to the right hand, since the vast majority of Auslan signs requires the right hand only, and in any case a right–handed Auslan speaker mostly uses his right hand, so the left channel is likely to contain less information than the right one.

Finally, a fourth choice has been to consider, for any frame, only the first six parameters related to hand positions and angles, since it is known that measurements about fingers are much less precise than the others.

With the above choices, the length for any database item reduces to $11 \cdot 6 = 66$ plus the class, and that for any DE individual drops down to $66 \cdot 13 = 858$. This might lead to a database with different features from the original one and thus to a classification with incorrect results, for example some signs might now be confused since useful information to distinguish them might have been lost, thus the experiments will also assess the goodness of the hypotheses made.

The reduced database is thus composed by a training set and a testing one. The former consists of 234 instances, divided into thirteen classes, one for each of thirteen signs listed above. Any class is represented in this training set by exactly 18 instances, i.e. those recorded in the first six weeks. The testing set, instead, consists of 117 elements, i.e. those in the last three weeks. As a result, the training set is assigned 67% of the database instances, and the testing set the remaining 33%, which are quite usual percentages in classification.

5.2   The Experiments

As concerns the other classification techniques used for the comparison we have made reference to the Waikato Environment for Knowledge Analysis (WEKA) system release 3.4 [10] which contains a large number of such techniques, divided into groups (bayesian, function–based, lazy, meta–techniques, tree–based, rule–based, other) on the basis of the underlying working mechanism. From each such group we have chosen some among the most widely used representatives. They are: among the bayesian the Bayes Net [11], among the function–based the MultiLayer Perceptron Artificial Neural Network (MLP) [12], among the lazy

Table 1. Achieved results in terms of $\%err$ and σ

	DE–C	BAYES NET	MLP ANN	IB1	KSTAR	BAG- GING	J48	NB TREE	PART	RIDOR	VFI
$\%err$	11.32	12.41	7.46	26.72	38.09	15.34	35.55	25.00	30.12	27.66	37.62
$\%err_b$	9.40	10.34	6.03	–	30.17	12.06	28.44	–	22.41	18.10	30.17
$\%err_w$	13.67	12.93	9.62	–	43.10	18.10	51.72	–	37.06	37.93	39.65
σ	1.17	0.90	1.30	–	4.43	1.73	7.07	–	4.08	4.62	2.11

IB1 [13] and KStar [14], among the meta–techniques the Bagging [15], among the tree–based ones J48 [16] and Naive Bayes Tree (NBTree) [17], among the rule–based ones PART [18] and Ripple Down Rule (Ridor) [19] and among the other the Voting Feature Interval (VFI) [20].

DE–C parameters have been set as follows: $n = 500$, $g = 3000$, $CR = 0.01$ and $F = 0.01$. The results are averaged over 20 runs differing one another only for the starting seed provided in input to the random number generator. For the other techniques, instead, some (MLP, Bagging, Ridor and PART) are based on a starting seed so that also for them 20 runs have been carried out by varying this value. Other techniques (Bayes Net, KStar, J48, VFI) do not depend on any starting seed, so 20 runs have been executed as a function of a parameter typical of the technique (*alpha* for Bayes Net, *globalBlend* for KStar, *numFolds* for J48 and *bias* for VFI). NBTree and IB1 depend neither on an initial seed nor on any parameter, so they have been executed only once.

DE–C execution time is around 5 minutes per run on a laptop with a 1.6–GHz Centrino processor. Thus times are comparable with those of the other techniques, which range from 5 seconds up to about 7 minutes for the MLP.

Table 1 shows the results achieved by the 11 techniques on the database. Namely, for any technique the average values of $\%err$ and the related standard deviations σ are given together with the best ($\%err_b$) and the worst ($\%err_w$) values achieved in the 20 runs. Of course σ is meaningless for NBTree and IB1. As it can be observed from the values in Table 1, DE–C is the second best technique in terms of $\%err$, very close to MLP which is the best and quite closely followed by Bayes Net. Bagging follows at a distance while all other techniques are quite far from the three best ones. The standard deviation σ for DE–C is not too high, meaning that, independently of the different initial populations, the final classifications achieved have similar correctness. Some techniques like J48, Ridor, KStar and Part, instead, show very different final values of $\%err$, thus sensitivity to different initial conditions.

Thus, the exploitation of DE to find positions of centroids has proven effective to face this reduced version of Auslan database: about 89 signs out of 100 are correctly recognized, on average, and 91 in the best case.

It should also be remarked that all the aforementioned choices aiming at reducing the amount of information have not caused the reduced database to be senseless: in fact all the techniques, each with its own performance, are able

Table 2. Confusion matrix for the best run effected by DE–C

	cold	drink	eat	exit	hello	hot	no	read	right	wait	write	wrong	yes
cold	9												
drink		9											
eat			9										
exit				9									
hello					9								
hot			1	1		7							
no							9						
read							7	1		1			
right							1		7				1
wait										9			
write											9		
wrong							3					6	
yes							1			1			7

to distinguish among the different hand signs. So those simplifying assumptions have proven useful and have not yielded negative consequences.

Table 2 shows the confusion matrix for the best run effected by DE–C, i.e. that with the lower $\%err$ on the test set. The row for the i–th class shows how its 9 examples have been scattered among the different classes (columns). The numbers on the main diagonal represent how many examples of any class have been correctly assigned to it. As it can be seen, 8 signs, like for instance *cold*, *drink* and *eat*, are in all cases recognized by the tool, whereas some others, such as *hot* and *yes*, are more difficult to distinguish. The sign *wrong* is in 3 cases confused with *no*. It is interesting to note that also in the best run of MLP *read* is confused once with *write* and *yes* is confused twice with *wait*, so this might depend on the way those signs were made, perhaps slower or faster or with very different hand position or rotation with respect to the "average" movement. Thus, those items might probably represent *outliers*.

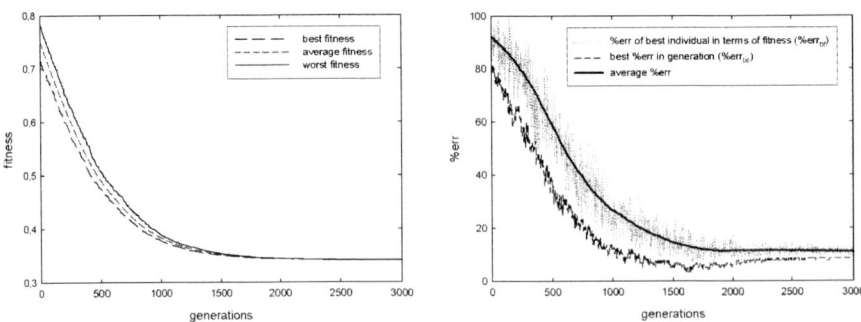

Fig. 1. Fitness (left) and $\%err$ (right) as a function of the generations for the best run

From an evolutionary point of view, in Fig. 1 we report the behavior of the best run in terms of lower final fitness value achieved. Its left part shows the evolution in terms of best individual fitness, average fitness and worst fitness in the population as a function of the number of generations. DE–C shows a first phase of about 700 generations in which fitness decrease is strong and almost linear, starting from 0.71 for the best and 0.75 for the average, and reaching about 0.42 for the best and 0.43 for the average. A second phase follows, lasting until about generation 1800, in which decrease is slower, and the two values tend to become closer, until they reach 0.3450 and 0.3456 respectively. From now on the decrease in fitness is slower and slower, about linear again but with a much lower slope, and those two values become more and more similar. Finally, at generation 3000 the two values are 0.3409 and 0.3410 respectively. The right part of Fig. 1, instead, reports the behavior of $\%err$ as a function of the generations. Namely, for any generation its average value, the lowest error value in the generation $\%err_{be}$ and the value of the individual with the best fitness value $\%err_{bf}$ are reported. This figure shows that actually the percentage of classification error on the testing set decreases as distance–based fitness values diminish, so confirming the hypothesis underlying our approach. It should be remarked here that $\%err_{bf}$ does not, in general, coincide with $\%err_{be}$, and is usually greater than it. This is due to the fact that our fitness does not take $\%err$ into account, so evolution is blind with respect to it, as it should be, and it does not know which individual has the best performance on the testing set.

What said above for a specific run is actually true for all the runs carried out, and is probably a consequence of the parameter setting used. This choice on the one hand allows a fast decrease in the first part of the run and, on the other hand, avoids the evolution being stuck in premature convergence as long as generations go by, as it is evidenced by the fact that best and average fitness values are different enough during the whole evolution.

6   Conclusions and Future Works

This paper has considered the issue of automatically recognizing digitally recorded human–made hand signs. To this aim, a reduced version of Auslan database has been accomplished and taken into account. A tool has been designed and implemented in which Differential Evolution is used to find the positions of the class centroids in the search space such that for any class the average distance of instances belonging to that class from the relative class centroid is minimized. The classification performance is estimated by computing the error percentage on the testing set for the resulting best individual.

The experimental results have proven that the tool is successful in tackling the task and is very competitive in terms of error percentage on the testing set when compared with other ten classification tools widely used in literature. In fact, only MLP has shown slightly better performance.

Future works will aim to further shed light on both the efficiency and the limitations of our evolutionary system: a plan is to endow DE–C with niching, aiming to investigate whether this helps in further improving performance.

References

1. Fayyad, U.M., Piatetsky-Shapiro, G., Smyth, P., Uthurusamy, R.: Advances in Knowledge Discovery and Data Mining. AAAI/MIT Press (1996)
2. Duda, R.O., Hart, P.E., Stork, D.G.: Pattern Classification. Wiley–Interscience, Chichester (2001)
3. Price, K., Storn, R.M., Lampinen, J.: Differential Evolution: A Pratical Approach to Globe Optimization. Springer, Berlin (2006)
4. Bäck, T.: Evolutionary Algorithms in Theory and Practice: Evolution Strategies, Evolutionary Programming, Genetic Algorithms. Oxford Univ. Press, Oxford (1996)
5. Johnston, T.A., Schembri, A.: Australian Sign Language (Auslan): An Introduction to Sign Language Linguistics. Cambridge University Press, Cambridge (2007)
6. Kadous, M.W.: Temporal Classification: Extending the Classification Paradigm to Multivariate Time Series, PhD Thesis, School of Computer Science and Engineering, University of New South Wales (2002)
7. http://kdd.ics.uci.edu/databases/auslan2/auslan.html
8. Blake, C.L., Merz, C.J.: UCI repository of machine learning databases, University of California, Irvine (1998),
   http://www.ics.uci.edu/tiny/~mlearn/MLRepository.html
9. Piater, J.H., Riseman, E.M., Utgoff, P.E.: Interactively training pixel classifiers. International Journal of Pattern Recognition and Artificial Intelligence 13(2), 171–194 (1999)
10. Witten, I.H., Frank, E.: Data Mining: Practical Machine Learning Tool and Technique with Java Implementation. Morgan Kaufmann, San Francisco (2000)
11. Jensen, F.: An Introduction to Bayesian Networks. UCL Press and Springer (1996)
12. Rumelhart, D.E., Hinton, G.E., Williams, R.J.: Learning representation by back-propagation errors. Nature 323, 533–536 (1986)
13. Aha, D., Kibler, D.: Instance–based learning algorithms. Machine Learning 6, 37–66 (1991)
14. Cleary, J. G., Trigg, L.E.: K*: an instance–based learner using an entropic distance measure. In: Proceedings of the 12th International Conference on Machine Learning, pp. 108–114 (1995)
15. Breiman, L.: Bagging predictors. Machine Learning 24(2), 123–140 (1996)
16. Quinlan, R.: C4.5: Programs for Machine Learning. Morgan Kaufmann Publishers, San Mateo (1993)
17. Kohavi, R.: Scaling up the accuracy of naive–bayes classifiers: a decision tree hybrid. In: Proceedings of the Second International Conference on Knowledge Discovery and Data Mining, pp. 202–207. AAAI Press, Menlo Park (1996)
18. Frank, E., Witten, I.H.: Generating accurate rule sets without global optimization. In: Machine Learning: Proceedings of the 15th International Conference, pp. 144–151. Morgan Kaufmann Publishers, San Francisco (1998)
19. Compton, P., Jansen, R.: Knowledge in context: a strategy for expert system maintenance. In: Proceedings of AI 1988, pp. 292–306. Springer, Berlin (1988)
20. Demiroz, G., Guvenir, H.A.: Classification by voting feature intervals. In: Proceedings of the 9th European Conference on Machine Learning, pp. 85–92 (1997)

## Part IV

Civil Engineering

Prediction of Ultimate Capacity of Laterally Loaded Piles in Clay: A Relevance Vector Machine Approach

Pijush Samui[1], Gautam Bhattacharya[2], and Deepankar Choudhury[3]

[1] Research Scholar, Department of Civil Engineering, Indian Institute of Science,
Bangalore - 560 012, India
pijush.phd@gmail.com
[2] Professor, Deparment of Civil Engineering, Bengal Engineering and Science University,
Howrah-711103, India
bhattacharyag@gmail.com
[3] Associate Professor, Department of Civil Engineering, Indian Institute of Technology
Bombay, Mumbai - 400076, India
dchoudhury@iitb.ac.in

Abstract. This study investigates the potential of Relevance Vector Machine (RVM)-based approach to predict the ultimate capacity of laterally loaded pile in clay. RVM is a sparse approximate Bayesian kernel method. It can be seen as a probabilistic version of support vector machine. It provides much sparser regressors without compromising performance, and kernel bases give a small but worthwhile improvement in performance. RVM model outperforms the two other models based on root-mean-square-error (RMSE) and mean-absolute-error (MAE) performance criteria. It also estimates the prediction variance. The results presented in this paper clearly highlight that the RVM is a robust tool for prediction of ultimate capacity of laterally loaded piles in clay.

Keywords: pile, clay, relevance vector machine, ultimate capacity.

1 Introduction

Piles are structural members made of timber, concrete, and/or steel, that are used for a variety of structures including heavy buildings, transmission lines, power stations, and highway structures. In many cases, piles are often subjected to considerable lateral forces such as wind loads in hurricane prone areas, earthquake loads in areas of seismic activity, and wave loads in offshore environments. So the determination of ultimate capacity of laterally loaded piles is an essential task in geotechnical engineering. There are several methods available for prediction of ultimate capacity (Q) of lateral loaded piles in clay ([1], [2], [3], [4]). Recently, artificial neural network (ANN) has been successfully used for pile analysis ([5],[6],[7],[8]).However, there are some limitations in using ANN. The limitations are listed below:

- Unlike other statistical models, ANN does not provide information about the relative importance of the various parameters [9].
- The knowledge acquired during the training of the model is stored in an implicit manner and it is very difficult to come up with reasonable interpretation of the overall structure of the network [10].

E. Avineri et al. (Eds.): Applications of Soft Computing, ASC 52, pp. 127–136.
springerlink.com   © Springer-Verlag Berlin Heidelberg 2009

- In addition, ANN has some inherent drawbacks such as slow convergence speed, less generalizing performance, arriving at local minimum and overfitting problems.
- Prediction is not probabilistic.

In this study, as an alternative method relevance vector machine (RVM) has been adopted to predict Q of lateral loaded pile in clay. RVM uses 44 load test data that have been taken form literature [11]. RVM is a probabilistic basis model. It relies on the Bayesian concept and utilizes an inductive modeling procedure that allows incorporation of prior knowledge in the estimation process [12]. The structure of the RVM model is identified parsimoniously. The main advantages of RVM are good generalization and sparse formulation. This paper has the following aims:

- To investigate the feasibility of the RVM model for predicting Q of laterally loaded pile in clay.
- To estimate the prediction variance.
- To compare the performance of RVM model with Broms (1964) [2] and Hansen (1961)[3] method.

2 Relevance Vector Machine

Tipping [13] proposed the RVM to recast the main ideas behind support vector machine (SVM) in a Bayesian context, and using mechanisms similar to Gaussian processes. A brief review of Tipping's paper is presented here for those unfamiliar with the work. The RVM model seeks to forecast y for any quarry x according to

$$y = f(x, w) + \varepsilon_n \qquad (1)$$

Where the error term $\varepsilon_n = N(0, \sigma^2)$ is a zero mean Gaussian process and $w = (w_0, ..., w_N)^T$ are a vector of weights. The likelihood of the complete data set can be written as

$$p(y/w, \sigma^2) = (2\pi\pi^2)^{-N/2} exp\left\{ -\frac{1}{2\sigma^2} \|y - \Phi w\|^2 \right\} \qquad (2)$$

Where $\Phi(x_i) = [1, K(x_i, x_1), K(x_i, x_2), ..., K(x_i, x_N)]^T$.

Without imposing the hyperparameters on the weights, w, the maximum likelihood of equation (2) will suffer from sever overfitting. Therefore Tipping (2001) [12] recommended imposition of some prior constraints, w, by adding a complexity penalty to the likelihood or the error function. An explicit zero-mean Gaussian prior probability distribution over the weights, w with diagonal covariance of α is proposed as follows:

$$p(w/\alpha) = \prod_{i=0}^{N} N\left(w_i / 0, \alpha_i^{-1}\right) \qquad (3)$$

With α is a vector of $N+1$ hyperparameters. Consequently, using Baye's rule, the posterior over all unknowns could be computed given the defined noninformative prior distribution:

$$p\left(w, \alpha, \sigma^2 / y\right) = \frac{p\left(y / w, \alpha, \sigma^2\right) p(w, \alpha, \sigma)}{\int p\left(y / w, \alpha, \sigma^2\right) p\left(w, \alpha, \sigma^2\right) dw\, d\alpha\, d\sigma^2} \tag{4}$$

Full analytical solution of this integral (4) is obdurate. Thus decomposition of the posterior according to $p\left(w, \alpha, \sigma^2 / y\right) = p\left(w / y, \alpha, \sigma^2\right) p\left(\alpha, \sigma^2 / y\right)$ is used to facilitate the solution [12]). The posterior distribution over the weights is thus given by:

$$p\left(w / y, \alpha, \sigma^2\right) = \frac{p\left(y / w, \sigma^2\right) p(w / \alpha)}{p\left(y / \alpha, \sigma^2\right)}$$

$$= (2\pi)^{-(N+1)/2} |\Sigma|^{-1/2} \exp\left\{-\frac{1}{2}(w-\mu)^T \Sigma^{-1}(w-\mu)\right\} \tag{5}$$

Where the posterior covariance and mean are respectively:

$$\Sigma = \left(\sigma^{-2} \Phi^T \Phi + A\right)^{-1}, \quad \mu = \sigma^{-2} \Sigma \Phi^T y \tag{6}$$

With $A = diag\left(\alpha_0, \alpha_1, ..., \alpha_N\right)$. Therefore learning becomes a search for the hyperparameter posterior most probable, i.e., the maximization of with respect to α and σ^2. For uniform hyperpriors over α and σ^2 one needs only maximize the term $p\left(y / \alpha, \sigma^2\right)$,

$$p\left(y / \alpha, \sigma^2\right) = \int p\left(y / w, \sigma^2\right) p(w / \alpha) dw$$

$$= \left((2\pi)^{\frac{-N}{2}} \middle/ \sqrt{\left|\sigma^2 + \Phi A^{-1} \Phi^T\right|}\right) \times \exp\left\{-\frac{1}{2} y^T \left(\sigma^2 + \Phi A^{-1} \Phi^T\right)^{-1} y\right\} \tag{7}$$

Maximization of this quantity is known as the type II maximum likelihood method ([14], [15]) or the "evidence for hyperparameter" [16]. Hyperparameter estimation is carried out in iterative formulae, e.g., gradient descent on the objective function [12]. The outcome of this optimization is that many elements of α go to infinity such that w will have only a few nonzero weights that will be considered as relevant vectors. The relevance vector can be viewed as counterparts to support vectors in SVM. Therefore the resulting model enjoys the properties of SVM and, in addition, provides, estimates of uncertainty bounds.

3 Analysis of RVM

In the present study, the above methodology has been used for prediction of Q of laterally loaded pile in clay. 44 laterally loaded pile test data were obtained from literature [11] and has been used in this study. The dataset contains information about the diameter of pile (D), embedded length of pile (L), load eccentricity (e), shear strength of clay (c_u) and Q. The parameters that have been selected are related to the geotechnical properties and the geometry of pile. More specifically, the input parameters are D, L, e and C_u and the output of model is Q. In carrying out the formulation, the data has been divided into two sub-sets: such as

(a) A training dataset: This is required to construct the model. In this study, 31 out of the 44 cases of pile load test are considered for training the dataset.
(b) A testing dataset: This is required to estimate the model performance. In this study, the remaining 13 data is considered as testing dataset.

The data is normalized against their maximum values [17]. To train the RVM model, two types of kernel function have been used: They are

- Gaussian
- Spline

In the present study, training and testing of RVM has been carried out using MATLAB [18].

4 Results and Discussions

In this study, the coefficient of correlation (R) is the main criterion that is used to evaluate the performance of the developed SVM models. Figures 1 and 2 illustrate the performance of the training dataset using Gaussian, and spline Kernel functions respectively, and the results are almost identical to the original data. In order to evaluate the capabilities of the RVM model, the model is validated with new data that are not part of the training dataset. Figures 3 and 4 shows the performance of the RVM model for testing dataset using Gaussian and spline kernel respectively. Table 1 shows the performance of RVM model for each kernel. Gaussian kernel exhibits slightly better performance than spline kernel. From the results, it is clear that the RVM model has predicted the actual value Q very well and it can be used as a practical tool for the determination of Q. Generally, good performance in the testing phase is considered to be evidence of an algorithm's practical plausibility and provides an evaluation of the model's predictive abilities. RVM has better performance in the training phase than in the testing phase. The loss of performance with respect to the testing set addresses a machine's susceptibility to overtraining. There is a marginal reduction in performance on the testing dataset (i.e., there is a difference between machine performance on training and testing) for the RVM model. So, RVM model has good generalization capability. The results have demonstrated that RVM is remarkable in producing an excellent generalization level while maintaining the sparsest structure. For example, the RVM model employs 25 to 36 percent of the training dataset as relevance vectors.

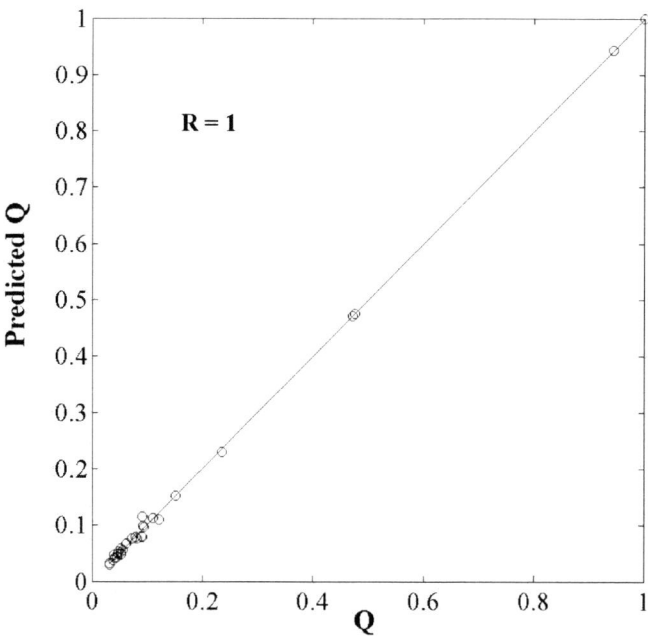

Fig. 1. Performance of RVM model for training dataset using Gaussian kernel

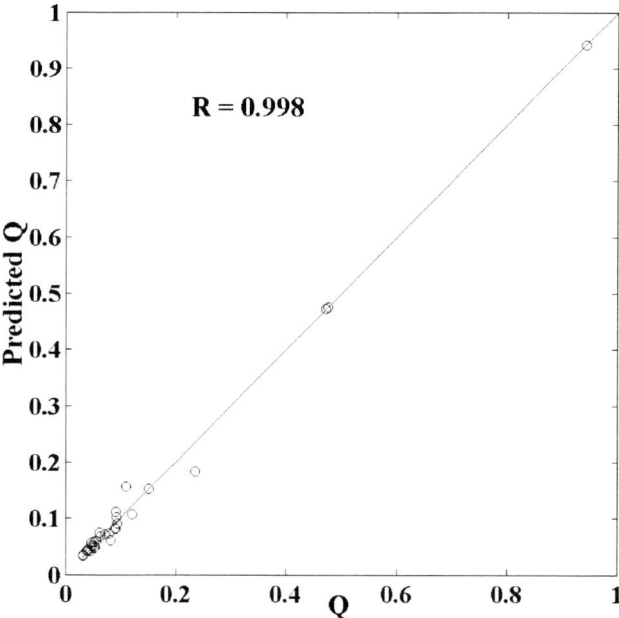

Fig. 2. Performance of RVM model for training dataset using spline kernel

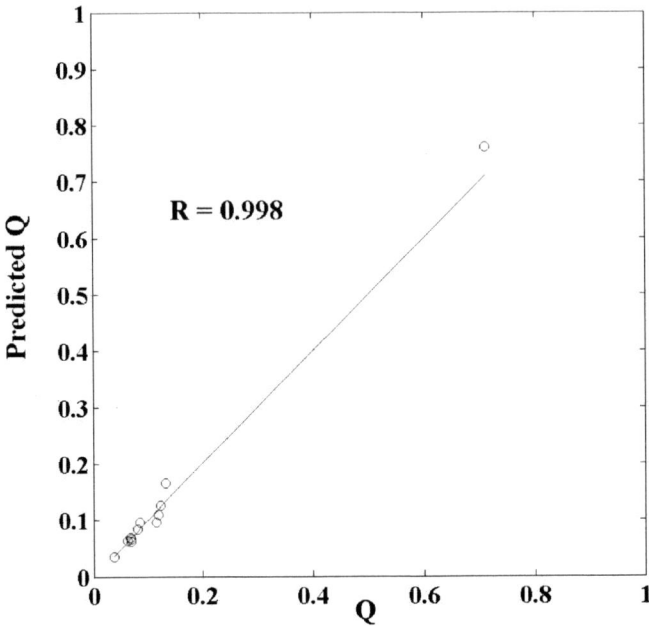

Fig. 3. Performance of RVM model for testing dataset using Gaussian kernel

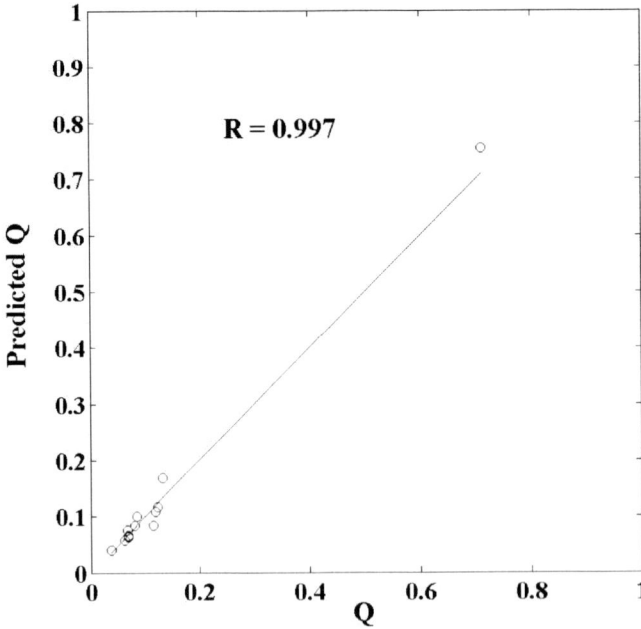

Fig. 4. Performance of RVM model for testing dataset using spline kernel

It is worth mentioning here that the relevance vectors in RVM represent prototypical examples [19]. The prototypical examples exhibit the essential features of the information content of the data, and thus are able to transform the input data into the specified targets. Figure 5 and 6 shows the prediction variance of testing dataset using Gaussian and spline kernel respectively.

The obtained prediction variance allows one to assign a confidence interval about the model prediction. A fundamental problem in machine learning occurs when one makes an attempt to infer parameters from a finite number of data points rather than from the entire distribution function. It is known that abundant data that accurately characterize the underlined distribution provide robustness for applications designed to specify model parameters. In forecasting models, there are infinitely many functions that may provide an accurate fit to the finite testing set. Notwithstanding this, RVM formulation does not try to fit data. Instead, it tries to capture underlying functions from which the data were generated irrespective of the presence of noise. For

Table 1. General performance of RVM for different kernels

Kernel	Training performance(R)	Testing performance(R)	Number of relevance vector
Gaussian, width= 0.4	1	0.998	11
Spline	0.998	0.997	8

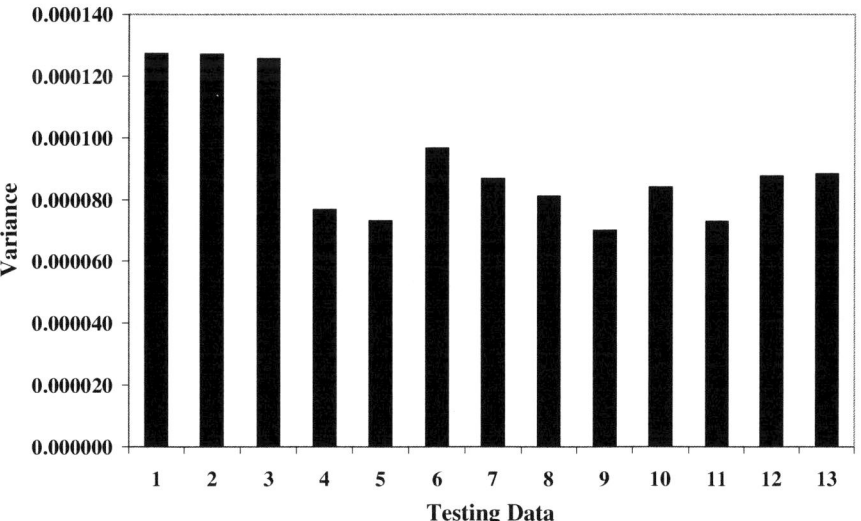

Fig. 5. Variance of testing dataset for Gaussian kernel

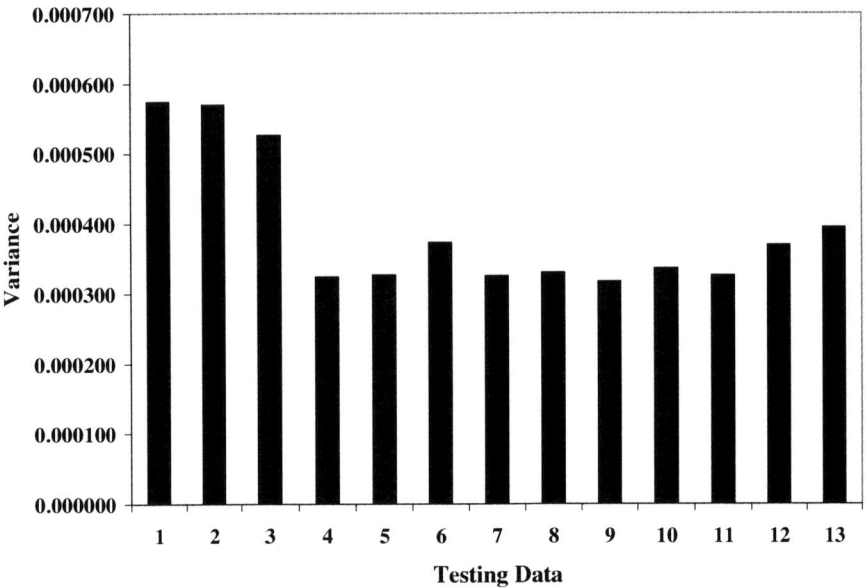

Fig. 6. Variance of testing dataset for spline kernel

RVM, the imposition of hyperparameters, i.e., maximization of the type-II likelihood, removes much of the noise in the data. In RVM, this feature also provides control over model complexity in ways that alleviate the problems of overfitting and underfitting.

The results for testing dataset have been compared with those available in the literature [11] which Q has been predicted by Broms and Hansen method. Table 2 represents the root-mean-square-error (RMSE) and mean-absolute-error (MAE) values for Broms, Hansen and RVM methods respectively. It is clear from Table 2 that a compared to Hansen method and Broms method such errors are less in the RVM based method.

Table 2. Comparison of RVM and Traditional Methods for Q prediction

Model	RMSE	MAE
RVM(Gaussian)	17.1674	11.0560
RVM(Spline)	18.3142	12.8870
Broms	34.6295	21.5977
Hansen	32.5781	30.9122

5 Conclusions

The RVM technique examined here have shown the ability to build model with high predictive capability for prediction of ultimate capacity of laterally loaded pile in clay. Data from 44 good-quality pile load tests in clay have been utilized to construct the model. RVM is characterized by its ability to represent the information content of the data set without being degraded in terms of model complexity by an abundance of data yet it is also computationally exhaustive during the training process. RVM is found to have excellent generalization properties, and have the added advantage of probabilistic interpretation that yields prediction uncertainty. The proposed RVM model is superior to the empirical ones when compared with actual measurements. In summary, this paper has helped to demonstrate that RVM can be viewed as powerful alternative approaches to physically based models for the prediction of ultimate capacity of laterally loaded pile in clay.

References

1. Czerniak, E.: Resistance to overturning of single, short piles. J. Structural Engg. 83(ST2), 1–25 (1957)
2. Hansen, B.J.: The ultimate resistance of rigid piles against transversal forces. Geo-tekenish institute Bull. No.12, Copenhagen (1961)
3. Broms, B.B.: Lateral resistance of pile in cohesive soils. J. Soil Mech. Found. Div. 90(SM2), 27–63 (1964)
4. Meyerhof, G.G.: The bearing capacity of rigid piles and pile groups under inclined loads in clay. Canadian Geotechnical Journal 18, 297–300 (1981)
5. Chow, Y.K., Chan, L.F., Liu, L.F., Lee, S.L.: Prediction of pile capacity from sress-wave measurements: A neural network approach. International Journal of Numerical and Analytical Methods in Geomechanics 19, 107–126 (1995)
6. Lee, I.M., Lee, J.H.: Prediction of pile bearing capacity using artificial neural network. Computers and Geotechnics 18(3), 189–200 (1996)
7. Abu Kiefa, M.A.: General Regression Neural Networks for driven piles in cohe-sionless soils. Journal of Geotechnical and Geoenviromental Engineering 124(12), 1177–1185 (1998)
8. Nawari, N.O., Liang, R., Nusairat, J.: Artificial Intelligence Techniques for the Design and Analysis of Deep Foundations. Electronics Journal of Geotechnical Engineering (1999)
9. Park, D., Rilett, L.R.: Forecasting freeway link ravel times with a multi-layer feed forward neural network. Computer Aided Civil And infa Structure Engineering 14, 358–367 (1999)
10. Kecman, V.: Leaning And Soft Computing: Support Vector Machines. In: Neural Networks, And Fuzzy Logic Models. MIT press, Cambridge (2001)
11. Rao, K.M., Suresh Kumar, V.: Measured and predicted response of laterally loaded piles. In: Proc. 6th International Conference and Exhibition on Piling and Deep Foundations, Bombay, pp. 1.6.1–1.6.7. Deep Foundation Institute (1996)
12. Tipping, M.E.: Sparse Bayesian learning and the relevance vector machine. J. Mach. Learn. 1, 211–244 (2001)
13. Tipping, M.E.: The relevance vector machine. Adv. Neural Inf. Process. Syst. 12, 625–658 (2000)
14. Berger, J.O.: Statistical Decision Theory and Bayesian Analysis, 2nd edn. Springer, New York (1985)
15. Wahaba, G.: A comparison of GCV and GML for choosing the smoothing parameters in the generalized spline-smoothing problem. Ann. Stat. 4, 1378–1402 (1985)
16. MacKay, D.J.: Bayesian methods for adaptive models. Ph.D. thesis, Dep. of Comput. and Neural Sysyt., Calif Inst. of Technol., Pasadena. Calif. (1992)

17. Sincero, A.P.: Predicting Mixing Power Using Artificial Neural Network. EWRI World Water and Environmental (2003)
18. MathWork Inc, Matlab user's manual, Version 5.3. Natick, MA, The MathWorks, Inc. (1999)
19. Li, Y., Campbell, C., Tipping, M.: Bayesian automatic relevance determination algorithms for classifying gene expression data. Bioinformatics 18(10), 1332–1339 (2002)

Air Quality Forecaster: Moving Window Based Neuro Models

S.V. Barai, A.K. Gupta, and Jayachandar Kodali

Department of Civil Engineering, Indian Institute of Technology
Kharagpur 721 30, India

Abstract. The present paper aims to demonstrate neural network based air quality forecaster, which can work with limited number of data sets and are robust enough to handle air pollutant concentrations data and meteorological data. Performance of neural network models is reported using novel approach of moving window concept for data modelling. The performance of model is checked with reference to other research work and found to be encouraging.

Keywords: Artificial neural networks, Moving window modeling, Forecasting, Maidstone.

1 Introduction

Air pollutants exert a wide range of impacts on biological, physical, and ecosystems. Their effects on human health are of particular concern. The decrease in the respiratory efficiency and impaired capability to transport oxygen through the blood caused by a high concentration of air pollutants may be hazardous to those who have pre-existing respiratory and coronary artery disease. The air-quality problems of a region may be characterized from two perspectives. (i) Violations of air-quality standards (ii) Public concern about air-quality; concern about human health effects; concern about visibility degradation and pollution colouration.

For such purposes, local environmental protection agencies are required to make air pollution forecasts for public advisories as well as for providing input to decision makers on pollution abatement and air quality management issues The present study aims at developing air quality forecasting models. The objectives of the study are as follows:

1. To collect multiple air quality parameters and meteorological   parameters for input and target data sets.
2. To identify the suitable inputs and targets for training the model.
3. To develop the neural network models for PM_{10}, SO_2, and NO_2
4. To validate the neural network models for PM_{10}, SO_2, and NO_2
5. To compare the results obtained by the neural network models and to evaluate the performance of the models with the help of statistical parameters.

2 Artificial Neural Networks and Air Quality Prediction

An artificial neural network is a massively parallel-distributed processor made up of simple processing units, which has a natural propensity for storing experiential

E. Avineri et al. (Eds.): Applications of Soft Computing, ASC 52, pp. 137–145.

knowledge and making it available for use (Haykin, 2001). Since the proposal of mathematical model for biological neuron by McCulloch and Pitts (1943), the artificial neural network development crossed the leaps and bounds and today many theories and training algorithms have come into practice. Backpropagation algorithm proposed by Rumelhart *et al.* (1986) is considered to be the best algorithm for function approximation, forecasting etc., and is generally used in combination with the multilayer feed forward networks (MLFF), also called as multilayer perceptrons (MLP). Multilayer perceptron networks consist of a system of simple interconnected neurons, or nodes representing a nonlinear mapping between inputs and outputs.

Although ANNs are relatively new, neural network models have proved to be a useful and very effective means for forecasting the pollutant concentrations (Pelliccioni and Poli, 2000; Kolehmainen *et al.*, 2000; Lu *et al.*, 2002; Lu *et al.*, 2003a, 2003b; Kukkonen *et al.*, 2003).

Ruiz-Suárez *et al.* (1995) used the Bidirectional Associative Memory (BAM) and the Holographic Associative Memory (HAM) neural network models for short-term ozone forecasting. Yi and Prybutok (1996) compared the neural network's performance for ozone concentration prediction with those of two traditional statistical models, regression, and Box-Jenkins ARIMA. Nunnari *et al.* (1998) used neural techniques for short and medium-range prediction of concentrations of O_3, nonmethyl hydrocarbons (NMHC), NO_2 and NO_x, which are typical of the photolytic cycle of nitrogen. Jorquera *et al.* (1998) compared different forecasting systems for daily maximum ozone levels at Santiago, Chile. The modelling tools used for these systems were linear time series, artificial neural networks and fuzzy models. Spellman (1999) estimated the summer surface ozone concentrations using surface meteorological variables as predictors by a multi-layer perceptron neural network for five locations in the UK. Gardner and Dorling (2000) compared linear regression, regression tree and multi-layer perceptron neural network (MLP) models of hourly surface ozone concentrations for U.K. data. Hadjiiski and Hopke (2000) developed ANN models to predict ambient ozone concentrations based on a limited number of measured hydrocarbon species, NO_x compounds, temperature, and radiant energy. Kao and Huang (2000) forecasted the SO_2, ozone concentrations using the conventional time-series approach and neural networks. The 1-step-ahead forecast gave better results than the 24-step-ahead forecast. Elkamel *et al.* (2001) developed an artificial neural network model that was able to predict ozone concentrations as a function of meteorological parameters and pollutant concentrations. Abdul-Wahab and Al-Alawi (2002) developed the neural network models to predict the tropospheric ozone concentrations as a function of meteorological conditions and various air quality parameters. Wang *et al.* (2003) developed an improved neural network model, which combines the adaptive radial basis function (ARBF) network with statistical characteristics of ozone in selected specific areas, and was used to predict the daily maximum ozone concentration level. Chaloulakou *et al.* (2003) have performed a comparison study with artificial neural networks (ANNs) and multiple linear regression models to forecast the next day's maximum hourly ozone concentration. Zolghadri *et al.* (2004) described the status of an on-going research program to develop a highly reliable operational public warning system for air pollution monitoring in Bordeaux, France. Ordieres *et al.* (2005) presented comparative studies on Multilayer Perceptron, Radial Basis Function and Square Multilayer perceptron for the prediction of average $PM_{2.5}$.

Corani (2005) compared the results of feed-forward neural networks, pruned neural networks and lazy learning models for air quality prediction in Milan. Shiva Nagendra and Khare (2005) studied vehicular exhaust emission neural networks models for predicting 8-hour average carbon monoxide concentration. Perez-Roa *et al.* (2006) developed model combining neural network and eddy diffusivity function K_v to predict peak concentrations of ambient carbon monoxide in a large city. Dutot *et al.* (2007) demonstrated neural classifier – Multilayer perceptron for a real time forecasting of hourly maximum ozon in the center of France.

Brief review of existing literature demonstrated tremendous potential of ANN application in air quality prediction problems. Following paragraphs will discuss about the study carried out for air pollution forecasting using novel moving model concept of data modelling.

3   Data Collection and Moving Window Model Deployment

3.1   Data Collection

Hourly air quality and meteorological data was collected from Kent Air Quality Monitoring Network (http://www.seiph.umds.ac.uk). The data of Maidstone from 16/04/2000(midnight) to 31/12/2001(11:00 P.M.) was used for the model development. Three pollutants were considered for the study purpose. They are respirable suspended particulate matter (PM_{10}), SO_2, and NO_2. Here, the meteorological parameters are wind speed, wind direction and temperature. The model was trained with 13480 training examples. The model's performance was tested for 1496 points.

3.2   Preprocessing

The data is normalized before giving as input to the network. In the present study normalizing is done by dividing the given data with the maximum value of the dataset.

3.3   Concept of Moving Window Model

A new methodology named, *moving window modeling* (Fig. 1) is used in developing all the models. A window is a subset of inputs and target(s) chosen along a set of time series data used for training.

The inputs and targets are named as the elements in a window. If elements of the window are pollutant concentrations (meteorological factors), then it is called as pollutant window (meteorological factor window). For e.g. a window containing PM_{10} concentration values as its elements can be called as PM_{10} window and a window containing temperature values can be called as a temperature window. The width of the window, no. of elements of the window and the time lag maintained between the chosen inputs, targets remain same. The time lag between two consecutive windows is equal to the minimum time interval of the time series, i.e. an hour for an hourly time series or a day for a daily time series. It is assumed that the window proceeds along the time series, learning the intricacies of the time series, during the training process,

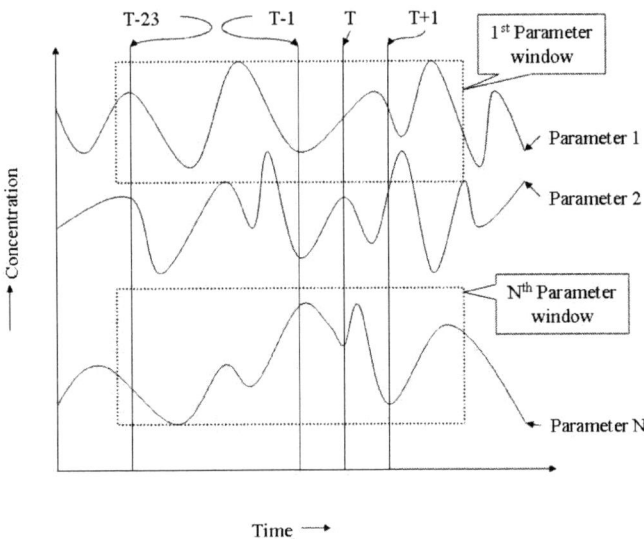

Fig. 1. Moving Window Concept

one step ahead every time as per the time lag. Each input factor follows its own moving window. Thus for considering p input factors, q target factors, we need to have p no. of moving windows with q no. of windows containing both inputs and the target.

3.4 Model Development

PM_{10}, SO_2, and NO_2 models with different inputs were developed using moving window concept discussed in previous section. It is assumed that historical values of the pollutant concentrations, meteorological parameters have a profound effect on the present pollutant concentration. Three pollutant parameters, namely PM_{10}, SO_2, NO_2 and three meteorological parameters, namely wind speed, wind direction and temperature are considered for model development. Pollutant concentrations, meteorological parameters at time T-23, T-1, T, T+1 are considered as the four elements of the respective windows as shown in Fig. .1. To take the effect of time into consideration three parameters, namely time of the day (T_{index}), day of the week (D_{index}), month of the year (M_{index}) are considered. T_{index} is assigned with values from 1 to 24 corresponding to 12:00 midnight and 11:00 P.M. D_{index} is assigned with values from 1 (Sunday) to 7 (Saturday). M_{index} is assigned with values from 1 (January) to 12 (December). All the input and target values are normalized between 0 and 1.

Time of the day, day of the week, month of the year at four time points (T-23, T-1, T, T+1) were the common inputs in all the forecast models as mentioned in the Table 1. In PM_{10} forecast model; PM_{10} concentrations at three time points (T-23, T-1, T), meteorological variables at all the four time points (T-23, T-1, T, T+1), were considered as additional inputs to the common inputs. PM_{10} concentration at T+1 time point was taken as the target or output. Same procedure is applied for SO_2 and NO_2 (M-1) models. In 2nd NO_2 forecast model (NO_2, M-2); NO_X concentrations at three time

Table 1. Inputs for various models

S.No.	Model	Inputs (time of the day, day of the week, month of the year at T-23, T-1, T, T+1 time points) and
1	PM_{10}	PM_{10} conc. at (T-23, T-1, T points), meteorological variables at (T-23, T-1, T, T+1 time points)
2	SO_2	SO_2 conc. at (T-23, T-1, T points), meteorological variables at (T-23, T-1, T, T+1 time points)
3	NO_2 (M-1)	NO_2 conc. at (T-23, T-1, T points), meteorological variables at (T-23, T-1, T, T+1 time points)
4	NO_2 (M-2)	NO_x conc. at (T-23, T-1, T+1, T points) additional to inputs for NO_2 (M-1)

points (T-23, T-1, T, T+1), were considered as additional inputs to (NO_2, M-1) model. NO_2 concentration at T+1 time point was taken as the target or output.

3.5 Neural Networks Implementation

The neural network models for this study were trained using the scaled conjugate gradient (SCG) algorithm (Moller, 1991). Training involves optimizing the network weights so as to enable the model to represent the underlying patterns in the training data. This is achieved by minimising the network error, for the training data set. The SCG algorithm has been shown to be very good improvement to standard back propagation and is also found to be faster than other conjugate gradient techniques (Moller, 1993). The learning algorithm used here was scaled conjugate gradient backpropagation of MATLAB neural network toolbox (Demuth and Beale, 2003). Two hidden layers were used along with an input and output layer. Networks with two hidden layers are more likely to escape poor local minima during the training process (Gardner and Dorling, 2000). Logarithmic sigmoidal transfer function was used in all the hidden, output layers and linear transfer function for input layer. The numbers of neurons in the hidden layers were determined by experimentation.

4   Results and Discussion

The performance of the models was measured with the help of correlation coefficient (R), mean absolute error (MAE) and index of agreement (d_2). Table 2 presents the performance statistics of the models in terms of above parameters when validated on an independent test data. Fig. 2 shows typical forecasting results for testing data of air pollutant- PM_{10}. The forecasted results are very close to observed parameters. Similar trend was observed for the other parameters too.

Further, there was a very good change of performance from NO_2 (M-1) to NO_2 (M-2). This can be attributed to the inclusion of historical values of NO_X concentrations as the extra model inputs.

Table 2. Statistical performances of PM_{10}, SO_2, and NO_2 models with varied inputs

S.No.	1	2	3	4
Model	PM_{10}	SO_2	NO_2 (M-1)	NO_2 (M-2)
Architecture	27-35-20-1	27-35-20-1	27-35-20-1	31-40-23-1
Epochs	7200	5800	5100	10000
Correlation coefficient (R)	0.8889	0.9189	0.8446	0.9654
Mean absolute error (MAE)	2.1054 $\mu g/m^3$	0.5678 ppb	3.4468 ppb	1.1932 ppb
Index of agreement (d_2)	0.938	0.957	0.913	0.982

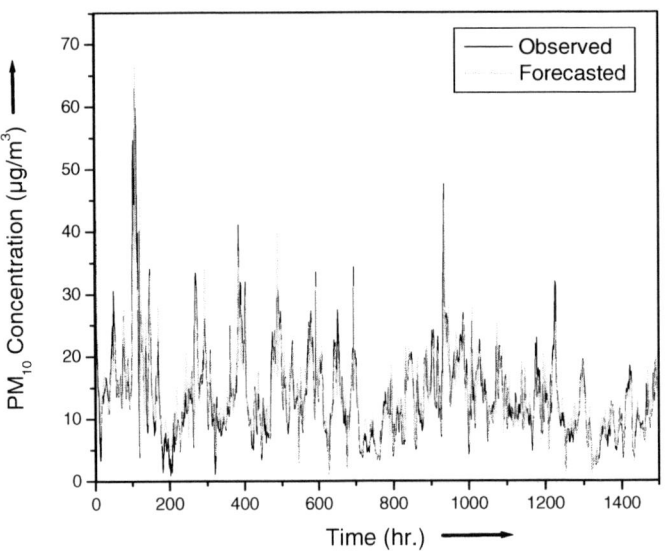

Fig. 2. Results of 1h ahead forecast for PM_{10} model

Kukkonen et al. (2003) reported the index of agreement ranging from 0.86 to 0.91 for NO_2. They however felt PM_{10} model's performance to be much lesser compared to NO_2. Kukkonen et al. (2003) has reported the index of agreement equal to 0.82

for NO_2 Kolehmainen et al. (2000) developed the MLP models for NO, NO_2, CO and PM_{10}. The index of agreement for the models developed found to be varying between 0.47 and 0.66. In the present case, index of agreement is found to be 0.938, 0.957, 0.913 and 0.982 for PM_{10}, SO_2 NO_2 (M-1) and NO_2 (M-2) respectively. The models thus developed are giving better predictions as compared to the existing neural network models.

5 Closing Remarks

In this paper, development of neural network based air quality forecast models using moving window concept is experimented successfully. The models thus developed possess some merits. Firstly, they are producing better results than the models developed by other reference results. Secondly, they are robust as seasonal factors are already considered in the model development.

References

Abdul-Wahab, S.A., Al-Alawi, S.M.: Assessment and prediction of tropospheric ozone concentration levels using artificial neural networks. Environmental Modelling and software 17, 219–228 (2002)

Chaloulakou, A., Saisana, M., Spyrellis, N.: Comparative assessment of neural networks and regression models for forecasting summertime ozone in Athens. The Science of the Total Environment 313, 1–13 (2003)

Corani, G.: Air quality prediction in Milan: feed-forward neural networks, pruned neural networks and lazy learning. Ecological Modelling 185, 513–529 (2005)

Demuth, H., Beale, M.: Neural network toolbox manual. The MathWorks, Inc. (2003)

Dutot, A., Rynkiewicz, J., Steiner, F.E., Rude, J.: A 24-hr forecast of ozone peaks and exceedance levels using neural classifiers and weather predictions. Environmental Modelling and Software 22, 1261–1269 (2007)

Elkamel, A., Abdul-Wahab, S., Bowhamra, W., Alper, E.: Measurement and prediction of ozone levels around a heavily industrialized area: a neural network approach. Advances in Environmental Research 5, 47–59 (2001)

Gardner, M.W., Dorling, S.R.: Statistical surface ozone models: an improved methodology to account for non-linear behaviour. Atmospheric Environment 34, 21–34 (2000)

Hadjiiski, L., Hopke, P.: Application of artificial neural networks to modeling and prediction of ambient ozone concentrations. J. Air & Waste Manage. Assoc. 50, 894–901 (2000)

Haykin, S.: Neural Networks: a Comprehensive Foundation, 2nd edn. Addison Wesley Longman, India (2001)

Jorquera, H., Pérez, R., Cipriano, A., Espejo, A., Letelier, M.V., Acuña, G.: Forecasting ozone daily maximum levels at Santiago, Chile. Atmospheric Environment 32(20), 3415–3424 (1998)

Kao, J.J., Huang, S.S.: Forecasts using neural network versus Box-Jenkins methodology for ambient air quality monitoring data. J. Air & Waste Manage. Assoc. 50, 219–226 (2000)

Kolehmainen, M., Martikainen, H., Hiltunen, T., Ruuskanen, J.: Forecasting air quality parameters using hybrid neural network modelling. Environmental Monitoring and Assessment 65, 277–286 (2000)

Kukkonen, J., Partanen, L., Karppinen, A., Ruuskanen, J., Junninen, H., Kolehmainen, M., Niska, H., Dorling, S., Chatterton, T., Foxall, R., Cawley, G.: Extensive evaluation of neural network models for the prediction of NO_2 and PM_{10} concentrations, compared with a deterministic modeling system and measurements in central Helsinki. Atmospheric Environment 37, 4539–4550 (2003)

Lu, W.Z., Fan, H.Y., Leung, A.Y.T., Wong, J.C.K.: Analysis of pollutant levels in central Hong Kong applying neural network method with particle swarm optimization. Environmental Monitoring and Assessment 79, 217–230 (2002)

Lu, W.Z., Wang, W.J., Wang, X.K., Xu, Z.B., Leung, A.Y.T.: Using improved neural network model to analyze RSP, NO_x and NO_2 levels in urban air in Mong Kok, Hong Kong. Environmental Monitoring and Assessment 87, 235–254 (2003a)

Lu, W.Z., Fan, H.Y., Lo, S.M.: Application of evolutionary neural network method in predicting pollutant levels in downtown area of Hong Kong. Neurocomputing 51, 387–400 (2003b)

McCulloch, W.S., Pitts, W.: A logical calculus of the ideas imminent in nervous activity. Bulletin of Mathematical Biophysics 5, 115–133 (1943)

Moller, M.F.: A scaled conjugate gradient algorithm for fast supervised learning. IEEE Transactions on Systems, Man and Cybernetics 21, 272–280 (1991)

Moller, M.F.: A scaled conjugate gradient algorithm for fast supervised learning. Neural Networks 6, 525–533 (1993)

Nunnari, G., Nucifora, A.F.M., Randieri, C.: The application of neural techniques to the modelling of time-series of atmospheric pollution data. Ecological Modelling 111, 187–205 (1998)

Ordieres, J.B., Vergara, E.P., Capuz, R.S., Salazar, R.E.: "Neural network prediction model for fine particulate matter (PM2.5) on the US-Mexico border in El Paso (Texas) and Ciudad Juarez (Chihuahua). Environmental Modelling 20, 547–559 (2005)

Pelliccioni, A., Poli, U.: Use of neural net models to forecast atmospheric pollution. Environmental Monitoring and Assessment 65, 297–304 (2000)

Perez-Roa, R., Castro, J., Jorquera, H., Perez-Correa, J.R., Vesovic, V.: Air-pollution modelling in an urban area: Correlating turbulence diffusion coefficients by means of an artificial neural network approach. Atmospheric Environment 40, 109–125 (2006)

Randerson, D.: A numerical experiment in simulating the transport of sulphur dioxide through the atmosphere. Atmospheric Environment 4, 615–632 (1970)

Rumelhart, D.E., Hinton, G.E., Williams, R.J.: Learning internal representations by error propagation. In: Parallel Distributed Processing 1, pp. 318–362. MIT Press, Cambridge (1986)

Ruiz-Suárez, J.C., Mayora-Ibarra, O.A., Torres-Jiménez, J., Ruiz-Suárez, L.G.: Short-term ozone forecasting by artificial neural networks. Advances in Engineering Software 23, 143–149 (1995)

Shiva Nagendra, S.M., Khare, M.: Modelling urban air quality using artificial neural network. Clean Technology Environment Policy 7, 116–126 (2005)

Spellman, G.: An application of artificial neural networks to the prediction of surface ozone concentrations in the United Kingdom. Applied Geography 19, 123–136 (1999)

Wang, W., Lu, W., Wang, X., Leung, A.Y.T.: Prediction of maximum daily ozone level using combined neural network and statistical characteristics. Environment International 29, 555–562 (2003)

Yi, J., Prybutok, V.R.: A neural network model forecasting for prediction of daily maximum ozone concentration in an industrialized urban area. Environmental pollution 92(3), 349–357 (1996)

Zolghadri, A., Monsion, M., Henry, D., Marchionini, C., Petrique, O.: Development of an operational model-based warning system for tropospheric ozone concentrations in Bordeaux, France. Environmental Modelling & Software 19, 369–382 (2004)

## Part V

**Computer Graphics,
Imaging and Vision**

Automatic 3D Modeling of Skulls by Scatter Search and Heuristic Features

Lucia Ballerini[1], Oscar Cordón[1], Sergio Damas[1], and José Santamaría[2]

[1] European Centre for Soft Computing, Edf. Científico Tecnológico, Mieres, Spain
{lucia.ballerini,oscar.cordon,sergio.damas}@softcomputing.es
[2] Dept. Software Engineering, University of Cádiz, Cádiz, Spain
jose.santamarialopez@uca.es

Summary. In this work we propose the use of heuristic features to guide an evolutionary approach for 3D range image registration in forensic anthropology. The aim of this study is assist to the forensic experts in one of their main tasks: the cranio-facial identification. This is usually done by the superimposition of the 3D model of the skull on a facial photograph of the missing person. In this paper we focus on the first stage: the automatic construction of an accurate model of the skull. Our method includes a pre-alignment stage, that uses a scatter-search based algorithm and a refinement stage. In this paper we propose a heuristic selection of the starting points used by the algorithm. Results are presented over a set of instances of real problems.

1 Introduction

Traditionally, the role of forensic anthropology has been the study of human remains mainly, for identification purposes [2]. The skull-photo superimposition method [6] is frequently used by anthropologists for the identification of an unknown skull because facial photographs are usually available, while other identification methods, as DNA analysis, are not always possible. During the superimposition process the 3D skull model is projected on the 2D image. In this paper we will focus our attention on the first stage of the process: the accurate construction of a model of the skull. Specifically, we focus in the frontal for the craniofacial identification. Our main goal is to provide forensics with an automatic and accurate alignment method.

There is a need to use a range scanner to develop a computerized study of the skull. This device is not able to cover the whole surface of the 3D object in just one scan. Hence multiple scans from different views are required to supply the information needed to construct the 3D model. The more accurate the alignment of the views, the better the reconstruction of the object. Therefore, it is fundamental to adopt a proper and robust *range image registration* (RIR) technique to align the views in a common coordinate frame, to avoid model distortion in the subsequent surface reconstruction stage.

Some range scanners are equipped with an additional positioning device named rotary table. It allows the proper calibration between scanner and the

E. Avineri et al. (Eds.): Applications of Soft Computing, ASC 52, pp. 149–158.
springerlink.com   © Springer-Verlag Berlin Heidelberg 2009

object while being rotated. Nevertheless, there are situations (corps and excavations) where it is not possible to use a rotary table, because it is not available or the size of the object to be scanned does not allow its use. The 3D skull model building when no positioning device is available is so complicated that it is one of the most time consuming tasks for the forensic experts. Existing software requires the manual and accurate selection of an important number (ten or more) of corresponding landmarks in the different views. Therefore, software tools for the automation of this work are a real need.

In a previous work we have proposed a method based on *evolutionary algorithms* (EAs) for the alignment of several range images acquired from skulls [10]. The method consists of a multi-stage approach: in the first stage, the *Scatter Search* (SS) EA [7] is used for pre-alignment; in the second stage a local optimizer is used for refinement, typically the *Iterative Closest Point* (ICP) algorithm [15]. Typically, in the pre-alignment step only a subset of points is used, while the refinement step is applied to the whole images.

In this work we propose a heuristic selection of points and compare results with the automatic approaches in [9]. The paper structure is as follows. In Section 2 we review our evolutionary-based method for the skull 3D model building. Section 3 introduces the heuristic procedure for point selection. The suitability of the method is tested in Section 4 over different scenarios of skull modeling. Finally, in Section 5 we present some conclusions and future works.

2   Range Image Registration

Our proposal is a feature-based approach to the RIR problem. The aim is finding a near-optimal geometric transformation that aligns every adjacent pair of views acquired by the range scanner. The general approach to this pair-wise RIR problem consists in two steps: pre-alignment and refinement. To do so, we use an efficient stochastic optimization technique named SS [7] for pre-alignment. This is a global search method, that provides a coarse alignment. Then a final refinement is applied, as a local search process, using the ICP algorithm [15]. This process is described in the following subsections.

2.1   Pre-Alignment

Basis of Scatter Search
SS fundamentals were originally proposed by Fred Glover in [4] and have been later developed in works [10, 7, 5]. The main idea of this technique is based on a *systematic* combination between solutions taken from a considerably reduced evolved pool of solutions named **reference set** (between five and ten times lower than usual GA population sizes). This way, an efficient and accurate search process is encouraged thanks to the latter and to other innovative components we will describe later. The general SS approach is graphically shown in Fig. 1. The fact that the mechanisms within SS are not restricted to a single uniform design allows the exploration of strategic possibilities that may prove effective in a particular implementation. Of the five methods in the SS methodology, only four are strictly

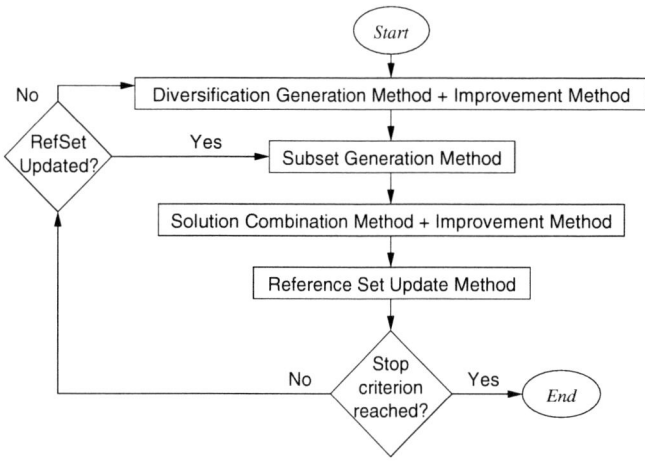

Fig. 1. The control diagram of SS

required. The *Improvement Method* is usually needed if high quality outcomes are desired, but a SS procedure can be implemented without it.

Coding Scheme and Objective Function

As coding scheme, the 3D rigid transformation f is determined by fixing seven parameters: rotation $R = (\theta,\ Axis_x, Axis_y, Axis_z)$ and translation $\mathbf{t} = (t_x, t_y, t_z)$, where θ and **Axis** define the 3D rotation given by an angle and an axis, respectively. Moreover, for a more suitable rotation representation, we consider quaternions instead of the three classical Euler matrices representation that suffers from the problem of *gimbal lock* [11].

In this contribution, we used the objective function introduced in our previous proposal [9]. It considers the *Median Square Error* to deal with the small overlapping between adjacent views:

$$F(f, I_s, I_m) \;=\; median\{\| f(\mathbf{p}_i) - \mathbf{p}_j' \|^2\}, \quad \forall \mathbf{p}_i \in I_s \qquad (1)$$

where f is the transformation we are searching for, I_s and I_m are the input scene and model views, $f(\mathbf{p}_i)$ is the point after the transformation is applied to the scene point $\mathbf{p}_i$, and $\mathbf{p}_j'$ is the closest model point to $f(\mathbf{p}_i)$ of the scene.

In order to speed up the closest point computation we consider an advanced data structure, the Grid Closest Point (GCP) [13]. We have performed an improvement using a Kd-tree structure [15] for the GCP cell initialization, instead of using the brute force search. So, GCP achieves the closest point search in constant time, except for those transformed scene points that fall outside the grid. In those cases, a Kd-tree search is performed.

SS-Based 3D RIR Implementation

Next, we will briefly describe the specific design of each component of our SS-based 3D IR method. For a detailed description of the different SS components, readers are referred to [10].

Diversification Generation Method: It makes use of a controlled randomization based on frequency memory to generate the initial set of diverse solutions.

Improvement Method: The Solis-Wetts algorithm is used as a local optimizer that generates trial solutions along the parameter transformation search space, guiding the optimization process to the next solution.

Subset Generation Method: It generates a collection of solution subsets of the Reference Set to create new trial solutions. We consider all the possible pairs of solutions.

Solution Combination Method: It is based on the use of the blend crossover operator (BLX-α) [3]. Such combination method was successfully incorporated to SS in [5]. The solution obtained by the BLX-α is selectively optimized by the *Improvement Method* and included in the *Pool* of solutions.

Reference Set Update Method: RefSet is updated to be composed of the b best solutions in $RefSet \cup Pool$ following a static strategy (first, the *Pool* set is built and then the updating is made) [7].

2.2 Refinement

Once we have a rough approximation to the global solution we will apply a local optimizer to refine the result from the first stage of the process. One of the most known feature-based algorithms for the refinement stage of RIR is the Iterative Closest Point (ICP), proposed by Besl and McKay [1] and extended in [15]. The algorithm has important drawbacks [1]: (i) it is sensitive to outliers; (ii) the initial states for global matching play a basic role for the method success when dealing with important deformations between model and scene points; (iii) the estimation of the initial states is not a trivial task, and (iv) the cost of a local adjustment can be important if a low percentage of occlusion is present. Hence, the algorithm performance is good only in the case of close initialization of the two images. As stated in [15]: "we assume the motion between the two frames is small or approximately known". This is a precondition of the algorithm to get reasonable results, and it holds when the pre-alignment achieves near-optimal results.

3   Heuristic Point Selection

In [9], we evaluated two point selection approaches: a semiautomatic and an automatic one. For the semiautomatic approach, the crest lines are extracted from the curvature information of the surface [14]. Points belonging to these lines are used as subset of features to reduce the large set of points involved. The crest lines provide a useful subset of data, because they have a very strong anatomical meaning (as pointed out by Subsol et al. [12], they emphasize the mandible, the orbits, the cheekbones or the temples), but their extraction requires the expertise of human intervention. The latter drawback is an obstacle for the anthropologists, because they do not have neither the knowledge nor the time to extract them. Moreover, the need for human intervention does not allow us to perform a fair comparison of this method with fully authomatic ones. On the other hand, an uniform random sampling of the input images is used in the automatic approach. A larger number of points needs to be selected in order to achieve a suitable accuracy, however it is fully automatic and really helpful for the anthropologists.

However, we observed that choosing points in a complete random way over the range images can produce non-representative samples that reside in areas having few geometric features and therefore this makes hard the identification of the corresponding point in the second image. Although it would be impossible to pick a truly representative sample, a heuristic can offer an improvement over the purely random approach. Let us see now some properties of the surfaces defined by the images acquired by the range scanner (see Fig. 2).

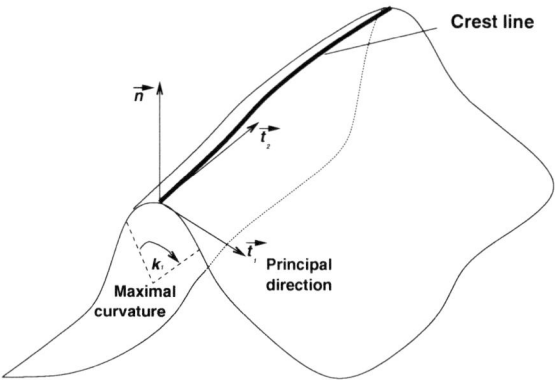

Fig. 2. Differential characteristics of surfaces

At each point P of those surfaces, there is an infinite number of curvatures but, for each direction $\mathbf{t}$ in the tangent plane at P, there is only one associated curvature $k_\mathbf{t}$. There are two privileged directions of the surface, called the principal directions ($\mathbf{t_1}$ and $\mathbf{t_2}$), which correspond to the two extremal values of the curvature: k_1 and k_2. One of these two principal curvatures is maximal in absolute value (let say k_1), and the two principal curvatures and directions suffice to determine any other curvature at point P. These differential values can be used in many different ways to locally characterize the surface. To those values, we can add the extremality criterion e as defined by Monga et al. in [8], which is the directional derivative of the maximal curvature (let say k_1), in the corresponding principal direction ($\mathbf{t_1}$). In fact, the same extremality criterion can be also defined for the other principal direction, and we therefore have two "extremalities" e_1 and e_2. The locations of the zero-crossing of the extremality criterion define lines, which are called ridge lines or crest lines.

For each point P we compute the curvature $k_\mathbf{t}$ using Yoshizawa et al. proposal [14]. We have chosen to discard all points having either very high or very low curvature, using two opportune threshold values selected according empirical experience. We discard points having low curvature, because they belong to flat areas and cannot give sufficient information to detect the corresponding point in the other mesh. We discard also points having high curvature, because they may belong to an open boundary (and thus to an incomplete local sampling of the surface) or could be due to noise generated during the acquisition process. The low threshold value is chosen in order to remove 10% of the total number of points, while the high one to remove 0.3% of them. These values, based on the observation of curvature values, has been selected empirically in our experiments and do not need to be tuned by the anthropologists. A uniform random sampling is then performed on the remaining points.

4   Experimental Study

In this section we aim to analyze our evolutionary approach to generate 3D skull models of forensic objects. We tackle the different problems the forensic expert has to deal with during the reconstruction stage of the photographic supra-projection process. Next, Section 4.1 describes the considered dataset. Sections 4.2 details the experimental design and the parameter settings. Finally, Section 4.3 is devoted to the analysis of results.

4.1   Input Range Images

The Physical Anthropology Lab of the University of Granada provided us with a dataset of human skulls acquired by a Konica-Minolta© 3D Lasserscanner VI-910. We focused our study on the range images of a the skull of a person who donated his remains for scientific proposes. The acquisition process includes noise removal and the use of smoothing filters.

We have taken into account important factors along the scanning process like time and storage demand. We considered a scan every $45°$, that is a resonable trade-off between number of views and overlapping regions. Hence, we deal with a sequence of eight different views: $0° - 45° - 90° - 135° - 180° - 225° - 270° - 315°$. The dataset we used in our experiments is only limited to five of them: $270° - 315° - 0° - 45° - 90°$. These five views allow to achieve a 3D model of the most interesting parts (the frontal ones) of the skull for the forensic expert and for the final objective of our work: the cranio-facial identification of a missing person.

4.2   Experimental Design and Parameter Settings

We focus our attention on the impact of the heuristic selection of points compared to the uniform random sampling used in [9]. We propose a set of RIR problem instances that simulate an unsupervised scanning process, i.e. not oriented by any device.

The original $270° - 315° - 0° - 45° - 90°$ views comprise 109936, 76794, 68751, 91590, and 104441 points, respectively. In the purely random RIR approach, we have followed an uniform random sampling of the input images. Hence, the only parameter the forensic expert must consider is the density of this random sampling. We fixed a 15% of the original dataset as a suitable value for the

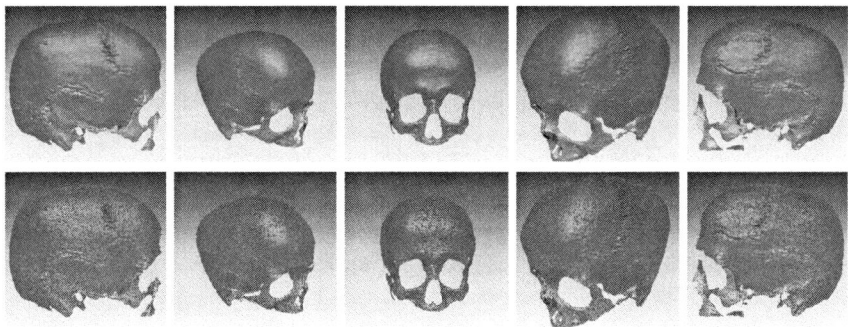

Fig. 3. From left to right: partial views of the skull and their corresponding selected points acquired at $270°, 315°, 0°, 45°$, and $90°$, respectively

time and accuracy trade-off. The heuristic approach is also fully automatic and once the low- and high-curvature points have been discarded, the density of the sampling of the remaining points is fixed to 15% of the total number of points of the original dataset. In this way, we have exactly the same number of points as in the purely random approach. Figure 3 shows the partial views of the skull and their corresponding selected points.

Specifically, we consider four different rigid transformations. They are shown in Table 1 and represent a typical bad initialization of the pre-alignment step by a forensic expert. Therefore, we are simulating some of the worst starting scenarios. Any method that aims at becoming considered a good RIR technique for this real world forensic application has to overcome such a bad initialization properly. These four transformations are applied to every adjacent pair of images of the 3D skull model ($0°$ vs $45°$, $0°$ vs $315°$, $45°$ vs $90°$, $315°$ vs $270°$) leading to a global set of sixteen pair-wise RIR problem instances to be solved. Therefore, every method finally deal with forty eight RIR problem instances (sixteen for every RIR approach: semiautomatic, purely random and heuristic).

Our method is run on a PC with an Intel Pentium D820 (2 core 2.8 GHz) processor, 2 GB RAM, Linux CentOS. In order to avoid execution dependence, fifteen different runs in each pair-wise RIR problem instance have been performed. The initial diverse set P comprises $Psize = 30$ solutions and the $RefSet$ is composed of the $b = 8$ best ones of them. BLX-α is applied with $\alpha = 0.3$, while the *Improvement Method* is selectively applied during 80 evaluations each time. The

Table 1. Four rigid transformations considered

	θ	$Axis_x$	$Axis_y$	$Axis_z$	t_x	t_y	t_z
T_1	115.0°	-0.863868	0.259161	0.431934	-26.0	-15.5	-4.6
T_2	168.0°	0.676716	-0.290021	0.676716	6.0	5.5	-4.6
T_3	235.0°	-0.303046	-0.808122	0.505076	16.0	-5.5	-4.6
T_4	276.9°	-0.872872	0.436436	-0.218218	-12.0	5.5	-24.6

Table 2. Minimum (m), maximum (M), median (ν), and standard deviation (σ) MSE values when tackling every problem instance (T_1, T_2, T_3, and T_4) related to the four pair-wise RIR scenarios from the automatic approach with purely random and heuristic point selection. The best minimum (m) and median (ν) values are in bold font.

scenario		T_1 / T_3 m	M	ν	σ	T_2 / T_4 m	M	ν	σ
I_{0^o}	random	1.517	13897.547	8782.495	3550.694	1.491	12890.772	5268.905	3474.097
	heuristic	**1.388**	14539.195	**6697.453**	3642.979	1.491	6696.869	**1.499**	2839.134
Vs.									
	random	1.496	3456.643	**1.506**	1277.753	1.449	10354.301	**6786.182**	4094.412
$T_i(I_{45^o})$	heuristic	1.506	5421.893	2965.474	1987.130	1.499	10097.473	8159.817	3955.081
I_{0^o}	random	0.023	0.208	0.037	0.044	0.014	19.522	0.040	4.858
	heuristic	**0.017**	660.684	**0.025**	165.008	0.018	0.073	**0.029**	0.017
Vs.									
	random	**0.006**	0.864	0.065	0.275	**0.006**	0.062	0.033	0.013
$T_i(I_{315^o})$	heuristic	0.023	0.072	**0.029**	0.016	0.023	0.058	**0.028**	0.014
I_{45^o}	random	1.350	20032.004	6.252	9778.392	1.094	19716.736	1.602	7828.282
	heuristic	**1.117**	19895.064	**1.416**	8612.617	1.088	19347.045	**1.417**	7598.472
Vs.									
	random	**1.162**	19403.543	1.460	4839.669	**1.098**	2.633	1.620	0.355
$T_i(I_{90^o})$	heuristic	1.266	19381.029	**1.323**	4834.145	1.293	19500.768	**1.319**	6620.805
I_{315^o}	random	2.083	20674.756	3.060	8256.547	2.250	21569.088	3.157	5379.434
	heuristic	**1.918**	20700.924	3.490	9110.726	**1.155**	21589.932	**2.874**	5384.822
Vs.									
	random	2.163	4.397	3.080	0.527	2.564	4.314	3.200	0.593
$T_i(I_{270^o})$	heuristic	**1.571**	3.873	3.544	0.714	**1.985**	3.661	**2.312**	0.665

execution time for the pre-alignment method is 100 seconds for the two automatic RIR approaches (using purely random or heuristic point selection). The stop criterion for ICP refinement stage is a maximum number of 250 iterations, which proved to be high enough to guarantee the convergence of the algorithm.

4.3 Analysis of Results

We have used the rotary stage as a positional device to actually validate the results for every RIR method. Since a high quality pre-alignment is provided from the scanner's software, when this device is available and a very experienced user performs the scanning, a 3D model is also available and it can be considered as the ground truth for the problem. Therefore, we know the global optimum location of every point in advance by using this 3D model. We use the usual *Mean Square Error* (MSE) to measure the quality of the process, once the RIR method is finished. MSE is given by:

$$MSE = \sum_{i=1}^{r} ||f(\mathbf{x}_i) - \mathbf{x}_i'||^2/r \qquad (2)$$

where $f(\mathbf{x}_i)$ corresponds to the i-th scene point transformed by f (which is the result of our RIR method), r is the number of points in the scene image, and $\mathbf{x}_i'$ is the same scene point but now using the optimal transformation f^* obtained by the positional device. Therefore, both $\mathbf{x}_i$ and $\mathbf{x}_i'$ are the same point but its

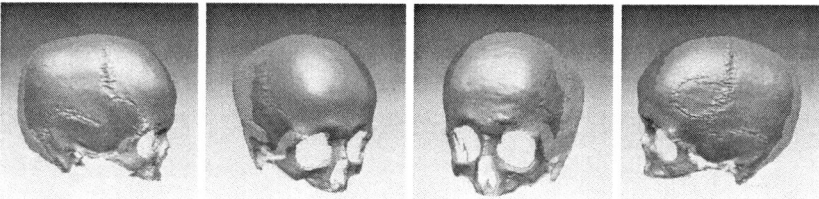

Fig. 4. The results of the four pair-wise RIR scenarios

location can differ if $f \neq f^*$. The availability of an *a priori* optimum model lets us to use this MSE definition to study the behavior of the RIR method in real world situations. Indeed, this evaluation is not applicable in real environments where no optimum model is available. The MSE is used by most authors and therefore it allows to compare our results with other methods.

Table 2 shows the results obtained with and without the use of heuristic information in the selection of points. Minimum (m), maximum (M), median (ν), and standard deviation (σ) MSE values over 15 runs are reported. The first conclusion is an increased performance. The heuristic selection reduced the error in fourteen of the sixteen instances. Specifically, the median (ν) values are smaller in twelve cases (three of them present a great improvement). It is worth to note that these results are obtained with the same number of points and exactly in the same execution time. It is also very important to note the robustness of SS in both approaches.

Finally, Fig. 4 aims to show some of our results. Specifically, we present one example for each of the four pair-wise RIR scenarios. Light and dark grey areas in every picture correspond to non overlapping regions of the views. Meanwhile, intermediate grey areas correspond to overlapping parts.

5   Concluding Remarks

We have detailed the suitability of range scanners for the reconstruction of reliable models in the forensic photographic supra-projection process. There are scenarios where a positional device which automatically builds the model cannot be used. Moreover, the positional devices often fail in building the 3D model. We have proposed an evolutionary methodology to solve the previous problems. Specifically, we have designed an experimentation of sixteen different pairwise RIR problems considering the 3D reconstruction of the skulls. From the results obtained, we have demonstrated that a fully automatic approach is possible using our SS-based proposal. The heuristic selection of starting points for the registration process improves the purely random sampling, while keeping the method fully automatic. We are planning to extend this study by including an automatic preprocessing stage (smoothing filter and noise removal) and a better design of the heuristic feature selection. Finally we aim at tackling the rest of the identification process, by the adaptation of our SS-based IR method to the 3D skull model-2D face photo superimposition, which is one of the most challenging tasks of the whole project.

Acknowledgments

This work was partially supported by the Spain's Ministerio de Educación y Ciencia (ref. TIN2006-00829) and by the Andalusian Dpto. de Innovación, Ciencia y Empresa (ref. TIC1619), both including EDRF fundings.

References

1. Besl, P.J., McKay, N.D.: Iterative point matching for registration of free-form curves and surfaces. IEEE Transactions on Pattern Analysis and Machine Intelligence 14, 239–256 (1992)
2. Cattaneo, C.: Forensic anthropology: development of a classical discipline in the new millennium. Forensic Science International 165(2-3), 185–193 (2007)
3. Eshelman, L.J.: Real-coded genetic algorithms and interval schemata. In: Whitley, L.D. (ed.) Foundations of Genetic Algorithms 2, pp. 187–202. Morgan Kaufmann, San Mateo (1993)
4. Glover, F.: Heuristic for integer programming using surrogate constraints. Decision Sciences, 156–166 (1977)
5. Herrera, F., Lozano, M., Molina, D.: Continuous scatter search: an analysis of the integration of some combination methods and improvement strategies. European Journal of Operational Research 169(2), 450–476 (2006)
6. Iscan, M.Y.: Introduction to techniques for photographic comparison. In: Iscan, M.Y., Helmer, R. (eds.) Forensic Analysis of the Skull, pp. 57–90. Wiley, Chichester (1993)
7. Laguna, M., Martí, R.: Scatter search: methodology and implementations in C. Kluwer Academic Publishers, Dordrecht (2003)
8. Monga, O., Benayoun, S., Faugeras, O.D.: Using partial derivatives of 3D images to extract typical surface features. In: Proceedings of the IEEE Computer Vision and Pattern Recognition (CVPR 1992), Illinois, USA, pp. 354–359 (1992)
9. Santamaría, J., Cordón, O., Damas, S.: Evolutionary approaches for automatic 3D modeling of skulls in forensic identification. In: Giacobini, M. (ed.) EvoWorkshops 2007. LNCS, vol. 4448, pp. 415–422. Springer, Heidelberg (2007)
10. Santamaría, J., Cordón, O., Damas, S., Alemán, I., Botella, M.: A scatter search-based technique for pair-wise 3D range image registration in forensic anthropology. Soft Computing 11(9), 819–828 (2007)
11. Shoemake, K.: Animating rotation with quaternion curves. In: ACM SIGGRAPH 1985, San Francisco, vol. 19, pp. 245–254 (1985)
12. Subsol, G., Thirion, J.-P., Ayache, N.: A scheme for automatically building three-dimensional morphometric anatomical atlases: application to a skull atlas. Medical Image Analysis 2(1), 37–60 (1998)
13. Yamany, S.M., Ahmed, M.N., Farag, A.A.: A new genetic-based technique for matching 3D curves and surfaces. Pattern Recognition 32, 1817–1820 (1999)
14. Yoshizawa, S., Belyaev, A., Seidel, H.: Fast and robust detection of crest lines on meshes. In: 2005 ACM Symp. on Solid and Physical Modeling, pp. 227–232 (2005)
15. Zhang, Z.: Iterative point matching for registration of free-form curves and surfaces. International Journal of Computer Vision 13(2), 119–152 (1994)

Multi-modal Registration Using a Combined Similarity Measure

Juan Wachs[1], Helman Stern[1], Tom Burks[2], and Victor Alchanatis[3]

[1] Department of Industrial Engineering and Management, Ben-Gurion University of the
  Negev, Be'er-Sheva, Israel, 84105
  {juan,helman}@bgu.ac.il
[2] Agricultural and Biological Engineering, University of Florida, Gainesville, FL, 110570
  TFBurks@ifas.ufl.edu
[3] Institute of Agricultural Engineering Volcani Center
  Bet-Dagan, Israel, 50250
  victor@volcani.agri.gov.il

Abstract. In this paper we compare similarity measures used for multi-modal registration, and suggest an approach that combines those measures in a way that the registration parameters are weighted according to the strength of each measure. The measures used are: (1) cross correlation normalized, (2) correlation coefficient, (3) correlation coefficient normalized, (4) the Bhattacharyya coefficient, and (5) the mutual information index. The approach is tested on fruit tree registration using multiple sensors (RGB and infra-red). The combination method finds the optimal transformation parameters for each new pair of images to be registered. The method uses a convex linear combination of weighted similarity measures in its objective function. In the future, we plan to use this methodology for an on-tree fruit recognition system in the scope of robotic fruit picking.

Keywords: Mutual information, multi-modal registration, similarity measures, sensor fusion.

1 Introduction

Multi-modal image registration is a fundamental step preceding detection and recognition in image processing pipelines used by the pattern recognition community. This preprocessing stage concerns the comparison of two images –the base and sensed images- acquired from the same scenario at different times or with different sensors in a way that every point in one image has a corresponding point on the other images, in order to align the images. This procedure has a broad use in the medical field to obtain insights regarding bone structure (CT scans) or tissues softness (MR scans), or to see the evolution of a patient based on images obtained over the years, see [1][2][3] for reviews in this field. Other fields that relies on multi-modal image registration preprocessing are remote sensing [4],[5], surveillance [6] and to mention a few. Our main problem is the automatic registration of fruit trees images obtained by multiple sensors. This is for the purpose of ultimately recognizing apples in a tree canopy using visual and thermal infrared cameras. In this paper, the initial problem of registering the fruit trees images by combining the images from different modalities is addressed. According to the position of the cameras with regard to the scene, it can be

E. Avineri et al. (Eds.): Applications of Soft Computing, ASC 52, pp. 159–168.

assumed that the transformation between images of different modalities are affine, which means rotations, translations, scaling and shearing are allowed. In this context, a standard image registration methodology called the correspondence problem includes the following steps [7]: a) Feature detection, b) Feature matching, and c) Transformation. Two approaches exist for feature detection step: a) Feature-based methods and b) Area-based methods. We focus in the area-based methods. The main representative in the area-based group was proposed by Viola and Wells [8] and is called mutual information methods, however a second group commonly used are the correlation-like methods [9].

A coarse-to-refined medical image registration method is presented in [10] by combining mutual information with shape information of the images. In [11] a new joint histogram estimation algorithm is proposed for computing mutual information to register multi-temporal remote sensing images, and is compared to correlation-like methods. Remotely sense image registration is addressed in [12] using the maximization of MI on a limited search space range obtained from a differential evolution strategy.

In this paper, an affine multi-modal registration method based on fusion and selection of SM is proposed. First, the sensed image is cropped to a "patch" and matched to areas in the base image. The best correspondence solution is achieved by maximizing the SM over the registration parameters search space. Solutions obtained by the SM are combined such that the mean squared error is minimized. We allow weighted use of all the similarity methods combined or the selection of the best method per affine parameter. Our approach is similar to [13] by adopting a combined methodology between powerful similarity measures, however we extend their work to include additional measures and to the affine registration scope.

The paper is organized as follows. A brief summary to mutual information and other similarity measures is given in section 2. In section 3, the proposed method is described in detail. Then we present experimental results in section 4, and finally conclusions in Section 5.

2 Mutual Information and Similarity Measures

Template matching methods are based on computing a SM between rectangular patches in an image pair. Corresponding patches between the two images are obtained when the maximum of the correlation is achieved. We will deploy three of these measures: cross correlation normalized (CC_1), correlation coefficient (CC_2), and correlation coefficient normalized (CC_3). The other two measures are histogram based: the Bhattacharyya coefficient (BC) and the Mutual Information index (MI). They rely on the joint/or and marginal histograms of the mutual areas. We denote V as a set of SM methods indexed as u = 1,2,3,4,5 for BC, MI, CC_1, CC_2 and CC_3, respectively.

2.1 Mutual Information (MI)

Let A, B be two random variables; let $p_A(a)$ and $p_B(b)$ be the marginal probability distributions; and let $p_{AB}(a,b)$ be the joint probability distribution.

The degree of dependence between A and B can be obtained by their mutual information (MI):

$$I(A,B) = \sum_{a,b} p_{AB}(a,b) \log \frac{p_{AB}(a,b)}{p_A(a) \cdot p_B(b)} \tag{1}$$

Given that H(A) and H(B) are the entropy of A and B, respectively, then their joint entropy is H(A,B); and H(A|B) and H(B|A) is the conditional entropy of A given B and B given A, respectively. Then, the entropy can be described by:

$$I(A,B) = H(A) + H(B) - H(A,B) \tag{2}$$

$$= H(A) - H(A|B) \tag{3}$$

$$= H(B) - H(B|A) \tag{4}$$

In terms of the marginal and joint probabilities distribution:

$$H(A) = -\sum_a p_A(a) \log p_A(a) \tag{5}$$

$$H(A,B) = -\sum_{a,b} p_{AB}(a,b) \log p_{AB}(a,b) \tag{6}$$

$$H(A|B) = -\sum_{a,b} p_{AB}(a|b) \log p_{A|B}(a|b) \tag{7}$$

In the context of registration, A is the sensed and B the base (or reference) images, and a and b are the grayscale value of the pixels in A and B, then the marginal and joint distributions $p_A(a)$, $p_B(b)$ and $p_{AB}(a,b)$ can be obtained by the normalization of the marginal and joint histograms of the overlapping areas in A and B. I(A,B) is maximum when the overlapping areas in A and B are geometrically aligned.

2.2 Histogram Comparison Using the Bhattacharyya Coefficient

In statistics, the Bhattacharyya distance measures the similarity of two discrete probability distributions. This attribute can be used to measure the similarity between two overlapping areas (A' $\subseteq$ A, B' $\subseteq$ B) in the sensed and base images.

BC(A',B') is maximum when the areas A' and B' are geometrically aligned.

$$BC(A',B') = \sum_{a \in A, b \in B} \sqrt{p_A(a) p_B(b)} \tag{8}$$

2.3 Correlation-Like Measures

Similarity measures (SM) are computed for pairs of overlapping areas between the sensed and base images. The maximum of these measures indicates corresponding areas. Let I_1 be the base image, I_2 be a patch in the sensed image with size (w ,h). Let i,j be coordinates in the overlapping area, then

The cross correlation normalized is:

$$CC_1(i,j) = \sum_{i',j' \in T} \frac{I_2(i',j') \cdot I_1(i+i',j+j')}{\sqrt{\sum_{i',j' \in T} I_2'(i,j)^2 \sum_{i',j' \in T} I_1'(i+i',j+j')^2}} \tag{9}$$

The cross correlation coefficient

$$CC_2(i, j) = \sum_{i', j' \in T} I_2'(i', j') \cdot I_1'(i+i', j+j')$$

where

$$I_2'(i'\, j') = I_2(i', j') - 1 \bigg/ \left[(w \cdot h) \cdot \sum_{i'', j'' \in T} I_2(i'', j'') \right] \qquad (10)$$

$$I_1'(i+i', j+j') = I_1(i+i', j+j') - 1 \bigg/ \left[(w \cdot h) \sum_{i'', j''} I_1(i+i'', j+j'') \right]$$

The cross correlation coefficient normalized

$$CC_3(i, j) = I_2'(i', j') \cdot I_1'(i+i', j+j') \bigg/ \sqrt{\sum_{i', j' \in T} I_2'(i, j)^2 \sum_{i', j' \in T} I_1'(i+i', j+j')^2} \qquad (11)$$

Some of the limitations of the CC based methods are: a) they are not able to cope with pairs of images where the sensed (template) and the base images differ by more than slight rotation and scaling. b) they cannot represent the intensity dependence between images from different modalities. Nevertheless, their simplicity and low time complexity compared to the MI method makes them useful for real-time applications.

3   Methodology

3.1   Transformation

Given that we want to register two input images referred to as the, the base A and the sensed B images from different modalities, and assuming that the scene presented in A is totally contained in B we want to find a transformation based on rotations, translations and isometric scaling that transforms every point in A to a point in B. This means that there exists a geometric transformation T_α defined by the registration parameter α such that $A(x,y)$ is related to $B(x,y)$. The optimal parameter α^* indicates that the images A and B are correctly geometrically aligned.

In this paper, we have restricted T_α to a 2D affine transformation. This is expressed by the registration parameter vector α by including a scaling factor s, a rotation angle θ (measured in degrees) and two translation distances t_x and t_y (measured in pixels). Hence $\alpha_j = [\alpha_{1j}, ..., \alpha_{ij}, ..., \alpha_{nj}]$ where α_{ij}=represents the ith parameter for the jth pair of images ($\alpha_{1j}=s$, $\alpha_{2j}=\theta$, $\alpha_{3j}=t_x$ and $\alpha_{4j}=t_y$ for the jth pair of images to be registered).

Transformation of the coordinates P_A and P_B from the sensed image A to the base image B is given by:

$$(P_B - C_B) = sR(\theta) \cdot (P_B - C_B) + t$$

$$R(\theta) = \begin{pmatrix} cos(\theta) & sin(\theta) \\ -sin(\theta) & cos(\theta) \end{pmatrix} \quad t = \begin{pmatrix} t_x \\ t_y \end{pmatrix} \qquad (12)$$

Where C_A and C_B are the coordinates of the centers of the images, $R(\theta)$ is the rotation vector, and t is the translation vector.

Then, a SM based on measure m_μ tries to solve the registration problem by finding the optimal registration parameter α^* such that the m_μ is maximized:

$$\alpha^* = \arg\max_{\alpha} m_{\mu}(A, B, \alpha) \tag{13}$$

Solving (13) does not always result in an optimum registration parameter, and the level of success in the registration depends on the capability of each SM to capture the relationship between the mutual areas in the images. However, it is possible that some SM's are more suitable for a specific registration parameters than others. In this vein, we try to find the best combination of SM's such that the absolute error between the optimal registration parameter and the observed one is minimized. The optimal registration parameter can be found in advance by manual registration.

3.2 Algorithm

Preprocessing. Initially, a contrast-limited adaptive histogram equalization algorithm is applied to the infra-red (IR) images to enhance their contrast. Secondly, we crop the sensed image to make a patch that fully overlaps the base image. We discard k=10% of each of the four sides around the IR images. This value was found empirically. A smaller value may leave areas not overlapping, and a higher value may leave out useful information and cause miss-registration.

Training. Given a training set of images S={$(a_1,b_1),\ldots\ldots,(a_m, b_m)$} where pair (a_j, b_j) represents the jth pair of sensed and base images that need to be registered. Let (w_a,h_a) and (w_b,h_b) be the with and height for the sensed and base images respectively. Let Λ and R be the parameter range for scaling and rotation respectively.

To register the images in the set S, follow the steps below:

---

1. For each sample pair of images $(a_j,b_j) \in$ S. Set k=0.
    2. For each s=$\alpha_1 \in \Lambda$, scale a_j by factor s,
        3. For each $\theta= \alpha_2 \in \Phi$, rotate b_j by θ,
            4. For each coordinates α_3=x $\in$ [0,…,w_b-w_a] and α_4=y $\in$ [0,…,h_b-h_a]
                5. Position the image a_j such that it left top corner coincides with x,y.
                6. Compute ρ_k=$m_{\mu}(a_j, b_j, \alpha_k)$ for all μ=1,…,V.
                7. k=k+1,
            8. End
        9. End
    10. End
    11. Find $\alpha_{j\mu}$ = arg max ρ
    12. Set $\Omega_{j\mu}$= $\alpha_{j\mu}$
13. End

Algorithm 1. Registration Algorithm

This algorithm results in a matrix Ω where the entries $\alpha_{ij\mu}$ are registration parameter i, for the pair of images j, using the SM m_{μ}. Given the true parameters obtained from manual registration, we denote the error of registration as: $e_{ij\mu} = (\alpha_{ij\mu} - \alpha_{ij\mu}^*)$.

For a training set of size N, the root mean square error (RMS) for parameter i using m_{μ} is

$$RMS_{i\mu} = \frac{1}{N} \sum_{j=1\ldots N} \sqrt{(\alpha_{ij\mu} - \alpha_{ij\mu}^*)^2} \tag{14}$$

Therefore we can suggest a combined SM, CM with weight vector $\{w_{i1},...,w_{ik},...,w_{i\mu}\}$ where the weight w_{ik} is associated with k^{th} SM for the registration parameter i, such that the RMS of the new combined method is minimized for each parameter of the registration vector. This is formulated in the following problem:

$$Min\,Z = RMS_{i\mu} = \frac{1}{N}\sum_{j=1,...N}\sqrt{(\alpha_{ij\mu}\cdot w_{i\mu} - \alpha_{ij\mu}*)^2}\quad,\qquad \mu\in V, i\in U \qquad (15)$$

s.t.:

$$\sum_{\mu\in V}w_{\mu i} = 1\,,\quad i\in U \qquad (16)$$

$$w_{\mu i}\geq 0\,,\quad \mu\in V, i\in U \qquad (17)$$

The approach proposed above suggests that the each registration parameter can be corrected by some weight such that the RMS is minimized. This implies that indirectly we assign weights to each of the SM explored so they can 'collaborate' together towards an optimal registration. An alternative approach is the selective SM, noted by LM, based on constraining (17) to integers only, such that the registration parameters are only corrected by one SM, each time.

The equations (15-17) can be solved using different optimization methods; in this work we used a pattern search method for linearly constrained minimization [15].

Testing. For the testing set of images $S'=\{(a'_1,b'_1),.......,(a'_m, b'_m)\}$ where each pair (a'_j, b'_j) of images is to be registered. Repeat the registration algorithm 1, however after step 11 add the line: $\alpha_{j\mu} = \alpha_{j\mu}\cdot w_{\mu j}$.

The values of $w_{\mu j}$ can be floating point or integers according to the LM or CM, respectively. The testing performance is obtained using again the RMS measure over the number of testing samples.

4 Experiments

This section presents a comparison in the performances of five SM's in the context of 2D non-rigid registration using fruit tree images from different modalities: RGB and IR. The performance measures are: MI, BC, CC_1, CC_2 and CC_3 and two additional indicators introduced in this paper: a combined (CM) and the selective (LM) methods. The registration algorithm proposed was tested using 2 different datasets. Dataset 1 contains 24 pairs of color and infrared images captured from a 3-5 meters distance to the fruit trees, while the images in Dataset 2 were 18 pairs, obtained from a 8-10 meters distance. We use manual registration as a "ground truth" reference to validate the registration performance of the different SM for datasets 1 and 2. Table 1 lists the input parameters and the range of the registration parameters.

The images of datasets 1 and 2 were registered using the five SM's: MI, BC, CC_1, CC_2 and CC_3. In each case the direct search optimization approach was used with the range of parameters in Table 1, and in the order (s, θ, t_x, t_y). Let $\alpha^*=\{s^*, \theta^*, t_x^*, t_y^*\}$ be the optimal parameters obtained from the manual registration and the registration parameters error be $\Delta\alpha = \{\Delta s, \Delta\theta, \Delta t_x, \Delta t_y\}$. Table 2 shows the RMS using both datasets (N=42) together for each SM. Note, that the best parameters were obtained by

Table 1. Input Parameters

Parameter	Value	
w_b, h_b (RGB Width and Height)	2560, 1920	
w_a, h_a (IR Width and Height)	320, 240	
Scale range (Λ)	6.8→ 7.2; δ=0.05	5.2→ 5.6; δ =0.05
Angle range (Φ)	-3 → 3; δ=0.5	
Translation x range	W/α_1-w	
Translation y range	H/α_1-h	
Joint Histogram size (only for MI)	256 x 256	

the MI method, except for the rotation which cc_1 resulted in the best value. The minimum error is highlighted in bold letters). As a comparative example, the performances of each *SM* are illustrated in Figure 1, when registering the images (a_j, b_j) for j=1, for Dataset 1. Figure 1.(f) shows a close-up around the top right apple. The mutual information method was able to keep the whole shape of the apple.

For this example, the joint probability of the pair of images was plotted, see Figure 2. The axes x,y are the coordinates on the overlap area using the base image axes. The peak of the surface gives the solution for the MI method, where the second best is the cc_2 measure.

A second experiment was conducted for validation purposes using the k-fold cross validation method, where k=N. This time the new two measures were added to the evaluation, LM and CM and were compared to MI.

Table 2. Registration error obtained over Dataset 1+2

Measure	RMS			
	Δs	$\Delta \theta$	Δtx	Δty
BC	0.254	1.547	58	86
MI	**0.177**	1.650	**19**	**20**
cc_1	0.237	**1.487**	60	73
cc_2	0.204	1.748	31	39
cc_3	0.212	1.769	31	40

The RMS obtained from all the sessions are presented in Table 3. The values of w_i were obtained from (15-18) using a pattern direct search method. The integer values of w_i were determined by selecting the w_i* that minimized the training error in the session, and later, the same value was used in the testing session.

The results show that the training performance for the combined measure CM was better than MI and LM for the training session.. When testing, the CM showed better results for the scale parameter only, while the LM showed better results or equal than MI. Therefore the use of both the CM and LM together, present an improvement in the performance of the registration for all the four parameters compared to the MI method. This validation suggests that for the cohort of image types tested, the transformation parameters found can be generalized to register new images.

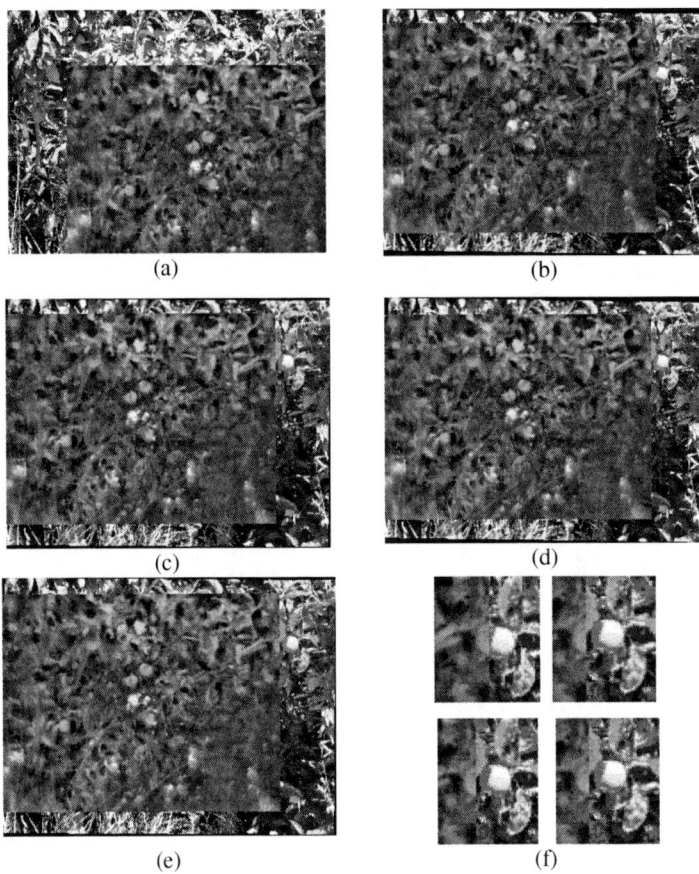

Fig. 1. Comparison of the five SM for a pair of images: (a) BC, (b) MI, (c) CC_1, (d) CC_2, (e) CC_3, (f) zoom-in in the top right apple for the last 4 methods. The top left which corresponds to (b) resulted in the best registration.

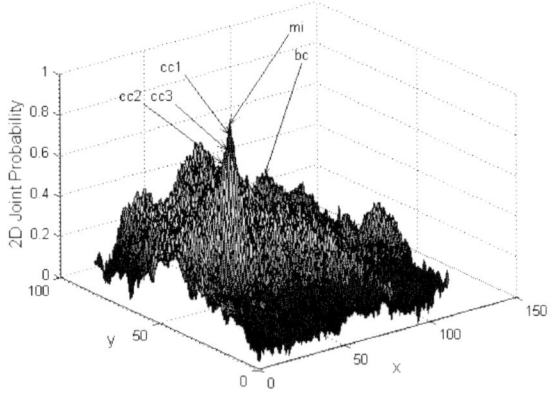

Fig. 2. Mutual Information Search Function and best solutions for each method

Table 3. Registration error obtained over Dataset 1+2 using k-fold cross with k=42

		MI		CM		LM	
		Train	*Test*	*Train*	*Test*	*Train*	*Test*
	Δs	0.177	0.177	**0.175**	**0.175**	0.177	0.177
RMS	Δθ	1.650	1.650	**1.265**	1.384	1.302	**1.303**
	Δtx	19	19	**17**	19	19	**19**
	Δty	20	20	**19**	21	20	**20**

4 Conclusions

In this work, we have presented a method of combining similarity measure of alternative image registration methods for multi-modal images registration. The measures used are: (1) cross correlation normalized, (2) correlation coefficient, (3) correlation coefficient normalized (CC3), (4) the Bhattacharyya coefficient, and the (5) the mutual information index. The registration parameters found are weighted according to the strength of each measure to predict each type of parameter, the combined approach (CM). An alternative approach is to find the best similarity measure per registration parameter (LM).

We found that the combined approach performed better than using each measure individually or using mutual information for the training sessions. During the testing session it was found that the prediction capability of the combination and best similarity measure approaches were better than that of using the mutual information measure. The method uses a convex linear combination of weighted similarity measures in its objective function.

In the future, we plan to use this methodology for on-tree fruit recognition system using multi-modal data, in the scope of robotic fruit picking.

Acknowledgments. This research was supported by Research Grant No US-3715-05 from BARD, The United States - Israel Binational Agricultural Research and Development Fund, and by the Paul Ivanier Center for Robotics Research & Production Management, Ben-Gurion University of the Negev.

References

1. Van den Elsen, Viergever, M.A.: Medical image matching - a review with classification. IEEE Engng. Med. Biol. 12, 26–39 (1993)
2. Maintz, J.B.A., Viergever, M.A.: A Survey of medical image registration. In: Spetzger, U., Stiehl, H.S., Gilsbach, J.M. (eds.) Navigated Brain Surgery, pp. 117–136. Verlag Mainz, Aachen (1999)
3. Lavallée, S.: Registration for Computer Integrated Surgery: Methodology, state of the art. In: Taylor, R.H., Lavallée, S., Burdea, G.C., Mosges, R.W. (eds.) Computer Integrated Surgery. MIT Press, Cambridge (1995)
4. Fan, X., Rhody, H., Saber, E.: Automatic Registration of Multi-Sensor Airborne Imagery. In: AIPR workshop, Washington, D.C, October 19-21 (2005)

5. Fransens, R., Strecha, C., Van Gool, L.: Multimodal and Multiband Image Registration using Mutual Information. In: Theory and Applications of Knowledge driven Image Information Mining, with focus on Earth Observation, EUSC, Madrid, Spain, March 17-18 (2004)
6. Krotosky, S.J., Trivedi, M.M.: Mutual Information Based Registration of Multimodal Stereo Videos for Person Tracking. Computer Vision and Image Understanding 106(2-3) (2007)
7. Zitova, B., Flusser, J.: Image Registration methods: a survey. Image and Computing 21, 977–1000 (2003)
8. Viola, P., Wells, W.M.: Alignment by maximization of mutual information. Intl. Journal of Computer Vision 24, 137–154 (1997)
9. Pratt, W.K.: Digital Image Processing, 2nd edn. Wiley, New York (1991)
10. Weiqing, C., Zongying, O., Weiwei, S.: A Coarse-to-Refined Approach of Medical Image Registration Based on Combining Mutual Information and Shape Information. In: Intl. Conference on Neural Networks and Brain, 2005. ICNN&B 2005, pp. 816–820 (2005)
11. Chen, H.-M., Varshney, P.K., Arora, M.K.: Performance of mutual information similarity measure for registration of multitemporal remote sensing images. IEEE Trans. on Geoscience and Remote Sensing 41(11), 2445–2454 (2003)
12. De Falco, D.C.A., Maisto, D., Tarantino, E.: Differential Evolution for the Registration of Remotely Sensed Images. Soft Computing in Industrial Applications, Recent and Emerging Methods and Techniques. In: Saad, A., Avineri, E., Dahal, K., Sarfraz, M., Roy, R. (eds.) Advances in Soft Computing, vol. 39 (2007)
13. Roche, A., Malandain, G., Ayache, N.: Unifying Maximum Likelihood Approaches in Medical Image Registration. Int. J. of Imaging Systems and Technology 11, 71 (2000)
14. Audet, C., Dennis Jr., J.E.: Analysis of Generalized Pattern Searches. SIAM Journal on Optimization 13(3), 889–903 (2003)

An Investigation into Neural Networks for the Detection of Exudates in Retinal Images

Gerald Schaefer and Edmond Leung

School of Engineering and Applied Science
Aston University
Birmingham, U.K
g.schaefer@aston.ac.uk

Abstract. We present an approach of automatically detecting exudates in retinal images using neural networks. Exudates are one of the early indicators of diabetic retinopathy which is known as one of the leading causes for blindness. A neural network is trained to classify whether small image windows are part of exudate areas or not. Furthermore, it is shown that a pre-processing step based on histogram specification in order to deal with varying lighting conditions greatly improves the recognition performance. Application of principal component analysis is used for dimensionality reduction and speed-up of the system. Experimental results were obtained on an image data set with known exudate locations and showed good classification performance with a sensitivity of 94.78% and a specificity of 94.29%.

Keywords: diabetic retinopathy, exudate detection, neural network, histogram equalisation/specification, principal component analysis.

1 Introduction

Diabetic retinopathy is a common eye disease directly associated with diabetes and one of the leading causes for blindness [1, 8]. Signs of the disease can be detected in images of the retina and manifest themselves, depending on the progression of the disease, as microaneurysms, intraretinal haemorrhages, hard exudates or retinal oedema. Hard exudates are one of the most common indicators of diabetic retinopathy and are formed in groups surrounding leakage of plasma. Over time, exduates increase in size and number; when they start to affect the macular region (the region of the retina that is used for sight focus and straight vision) this can eventually lead to blindness. Mass screening efforts are therefore underway in order to detect the disease at an early stage when it is still treatable. Unfortunately, these screening programmes are both cost and labour intensive as they require experts to manually inspect the captured retinal images. Automated techniques that are able to reduce the workload of the specialists are therefore highly sought after.

In this paper we introduce such an approach for automatically detecting exudates in retinal images. We employ a sliding window approach to extract subregions of an image. We then train a backpropagation neural network to distinguish

E. Avineri et al. (Eds.): Applications of Soft Computing, ASC 52, pp. 169–177.

exdudate locations from non-exudate areas. In order to deal with different lighting conditions commonly encountered in retinal images we modify the colour histograms of the images so as to match a common reference image. For dimensionality reduction we employ principal component analysis which in turn reduces the complexity of the neural net and hence improves training and classification speed. The proposed method is evaluated on a database of retinal images with known exuduate locations (marked by an expert). A sensitivity of 94.78% and a specificity of 94.29% on unseen images confirm the effectiveness of our approach.

2 Background

Gardner *et al.* [3] presented one of the earliest works on using neural nets for retinal image analysis and developed an algorithm that was able to identify vessels, exudates and haemorrhages. A retinal image was divided into disjoint 20×20 pixel regions and each region assigned by an expert as either exudates or non-exudate. Each pixel of the window corresponds to an input of a backpropagation network giving a total of 400 inputs. A sensitivity of 93.1% in detecting exudates was achieved.

The method by Osareh *et al.* [7] uses histogram specification [5] as a pre-processing step to eliminate colour variations and images are then segmented using a fuzzy c-means clustering technique. Each segmented region is classified as either exudate or non-exudate and characterised with 18 features. A two-layer perceptron network was trained with these and a sensitivity of 93% and specificity of 94.1% were reported.

Walter *et al.*'s [10] approach relies on morphological image processing to isolate exudate regions. First, candidate regions are found based on high local contrast variations while the exact contours of exudate regions are then extracted using morphological operators. They report a sensitivity of 92.8%.

Sinthanayothin *et al.* [9] used a recursive region growing technique that groups similar pixels together. Following a thresholding step, the resulting binary image shows the extracted exudate areas. Using this approach a sensitivity of 88.5% and specificity of 99.7% were achieved.

3 Neural Network Based Detection of Exudates

In our work the aim was to investigate how a relatively "naïve" neural network performs for exudate detection in retinal images. In contrast to other work we wanted to keep both pre-processing as well as the actual input features for the network as simple as possible. The general structure of our approach is shown in the flowchart of Figure 1. From there it can be observed that we evaluated several different configurations. The simplest of these is where the input to the neural network is taken directly from the retinal images, i.e. the raw image data. We then explore image pre-processing techniques, namely histogram equalisation and histogram specification to compensate for non-uniform lighting conditions

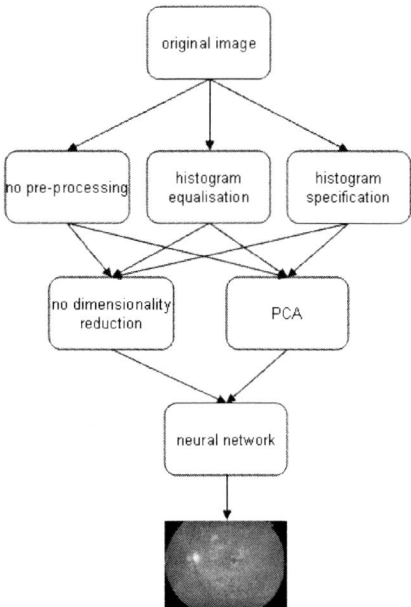

Fig. 1. Flowchart of our exudate detection approach taking into account different pre-processing and feature representation methods

present in the images. Finally, we also investigate the application of principal component analysis for dimensionality reduction and efficiency enhancement.

3.1   Input Features

The input features constitute the data the neural network is fed with in order to reach a decision, i.e. in order to detect exudates. Since, as stated above, we were interested in keeping these features as simple as possible we make use of the direct, i.e. raw image data and employ a sliding window technique where a mask of certain size is moved over the complete image so that each image pixel is once located at the centre of the mask - the image pixels covered by the mask then serve as the input features for that particular pixel. In contrast to other work which often focusses on the green channel only we make use of the full colour data (i.e. the red, green, and blue pixel intensities) within a retinal image. Various window sizes were evaluated and a final mask size of 9×9 pixels chosen so as to take into account the final size of the network inputs with respect to the feature vector and the amount of information to present to the network.

If the central pixel of a window is part of an exudate region then it is added as a positive sample to the training set. On the other hand, if the centre of the window is not in an exudate region, then this is regarded as a negative sample. A separate target vector is created to aid training, and depending whether the feature vector is positive or negative, set to either 0 or 1.

3.2   Training Data

It is evident that a neural network does not inherently "know" what exudates look like or how to detect them reliably. Therefore, as in all neural network applications the network has to be trained for the task where the training set must comprise both positive data covering exudates, and negative data of non-exudate regions. For this we employ a ground truth datset of 17 colour fundus images [10]. Each image has a resolution of 640 x 480 and is stored in 24-bit full colour format. All retinal images show some signs of diabetic retinopathy with candidate exudates and all exudates have been marked, to pixel resolution, by an expert.

Several strategies can be employed when training a neural network. In our work, we first (randomly) select 10 of the images whose extracted positive and negative samples are used for training. In order to ensure good generalisation ability on unseen data, early stopping is used to prevent over-fitting based on the training data. We use a validation set of 2 images (from the remaining 7 images) to perform the necessary validation. A 50-fold validation step is used, where after every 50 iterations during the training period the network is tested on the validation set. The current error rate is compared with the previous error and training is terminated when the validation error increases.

Compared to the background, the exudate regions are rather sparse. Therefore, if all the image data were extracted for training, the resulting network would be biased towards negative (i.e. non-exudate) samples. To avoid this we extract all positive samples from each image and a matching number of randomly located negative samples to end up with a training set of about 40,000 feature vectors.

Using the above 15-2-1 training scheme allows us to train a neural network with ability to generalise over unseen images. We also tested a second training strategy where we make use of the maximum possible amount of training data, that is generating a training set from 16 of the 17 images. As before, negative samples are extracted so as to match the number of positive training samples from the images. In order to evaluate the classification performance over the whole data set, training is performed 17 times, each time training on 16 images while testing on the left out case. Depending on the chosen images, training set sizes vary between 39,000 and 49,000 feature vectors.

3.3   Neural Network Structure

A simple fully-connected three-layer perceptron architecture with one input, one hidden, and one output layer was used for the neural network. The input layer consists of $9*9*3 = 243$ neurons, corresponding to the red, green, and blue pixel values of the 9×9 pixel area. We varied the number of neurons in the hidden layer, testing a range of different sizes before settling for 50 hidden neurons which provided a good trade-off between classification performance and network complexity. The output layer contains only a single neuron whose output is mapped, through a non-linear logistic output function, to the range of $[0; 1]$. Since a standard back-propagation algorithm with gradient descent to minimise the error during training would yield slow convergence [6], a more optimised variant of the algorithm, namely a scaled conjugate gradient method, was adopted as training paradigm.

When the neural network is used as a classifier, it is passed an unseen 9×9 image area to produce a network output in the $[0; 1]$ range. A threshold is then used to perform the final classification which determines whether the centre pixel of the input area is marked as part of an exudate or not.

3.4   Colour Normalisation

As it is difficult to control lighting conditions but also due to variations in racial background and iris pigmentation, retinal images usually exhibit large colour and contrast variations, both on global and local scales. Goatmana *et al.* [4] evaluated three pre-processing techniques on retinal images including greyworld normalisation, histogram equalisation [5, 2] and histogram specification [5] in order to reduce the variation in background colour among different images. They found that histogram specification performed best followed by histogram equalisation. In our experiments we also use these two techniques as a pre-processing step.

Histogram equalisation alters the histogram to approximate a uniform flat shape where the different intensities are equally likely. Histogram equalisation was applied to each of the colour channels red, green and blue, spreading the intensity values across the full range of the image. Histogram specification on the other hand, involves approximating the histogram shape of an image in correspondence to the desired histogram of another image. A target image was hence selected and all other images histogram specification applied to all other images based on this target.

3.5   Principal Component Analysis

As mentioned above, the neural network consists of 243 input neurons, 50 hidden neurons and 1 output neuron. As the network is fully connected this results in $243 * 50 + 50 = 12200$ weights that are being updated for each of the ca. 40,000 training samples during one training epoch which clearly constitutes a considerable computational burden during the training process. Training time grows exponentially with network size, and we have therefore investigated the application of principal component analysis (PCA) to reduce the number of input features and hence the computational complexity [6].

After performing PCA we select the minimal number of components required for representing at least 97% of the variance in the training data. For the case where we train the neural network on the raw image data this equates to 30 principal components, and hence an input layer with 30 neurons. Combined with a hidden layer with 15 neurons this means $30 * 15 + 15 = 465$ rather than the 12200 weights when using the full data and hence a network complexity reduced by a factor of more than 25.

For histogram equalised images 90 components are needed to represent 97% of the variance whereas for the images that undergo histogram specification the resulting size of the input layer is 50 neurons. The number of hidden neurons we used for these two cases was 40 and 20 respectively.

4   Experimental Results

As described above we perform two different sets of experiments corresponding to the two different training strategies employed. In the first test, we use 10 of the 17 images for training, 2 for validation and the remaining 5 for testing, while for the second test we follow a leave-on-out strategy training on 16 images and evaluating it on the remaining one. For both configurations we generate several neural networks: one where the input is the raw image data, one with histogram equalised images as input and one where the images underwent histogram specification before serving as input to the neural network. We also employ PCA on these different types of input data and hence create three further neural networks for this purpose to end up with 6 different network configurations.

In order to perform an objective evaluation of the achieved performance we compare the results of the classification provided by the various neural networks to the ground truth that was marked manually by a specialist. For each test image we record the number of True Positives TP (the number of pixels that were classified both by the network and the specialist as exudate pixels), True Negatives TN (the number of pixels that were classified both by the network and the specialist as non-exudate pixels), False Positives FP (the number of instances where a non-exudate pixel was falsely classified as part of an exudate by the neural network) and False Negatives FN (the number of instances where an exudate pixels was falsely classified as non-exudate by the neural network). From this we can then calulate the sensitivity SE (or true positive rate) as

$$SE = \frac{TP}{TP + FN} \tag{1}$$

and the specificity SP (or true negative rate) as

$$SP = \frac{TN}{TN + FP} \tag{2}$$

Furthermore, by varying the threshold of the output neuron we can put more emphasis on either sensitivity or specificity and also generate the ROC (receiver operator characteristics) curve which describes the performance of a classifier's sensitivity against its specificity and is used extensively in medical applications [11]. As a single measure of performance the area under the curve (AUC) can be used to compared different classifiers or different configurations. A perfect classification model identifying all positive and negative cases correctly would yield an AUC value of 1.0, whereas an AUC of 0.5 would signify random guessing.

Figure 2 on the left shows the ROC results of taking training samples from 10 images, using 2 images for validation and performing classification on the remaining 5 images. Using solely raw colour data from the original images as

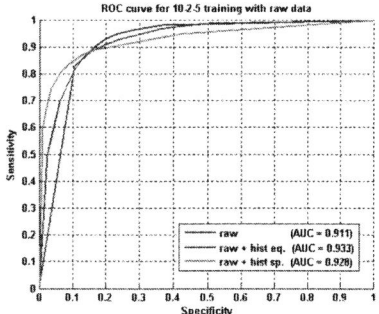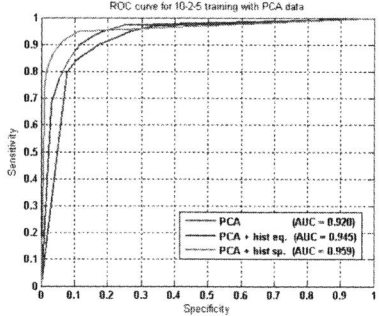

Fig. 2. ROC curves, based on 10-2-5 training, on raw data (left) and PCA data (right)

input to the neural network seems to obtain reasonable results, resulting in an AUC value of 0.911. Pre-processing the images with histogram equalisation before passing the data to the network allows for a higher detection rate of true positives. The resulting AUC value is 0.933, a clear improvement over that achieved by training on raw data. Using histogram specification performs fairly similar with an AUC value of 0.928.

Figure 2 on the right provides the results of the neural networks whose input patterns are generated from mapping the image data to the lower dimension space spanned by PCA components. Comparing these results with the ones obtained on raw data we can see that not only is training much more efficient due to the reduced complexity of the resulting networks but also the the achieved accuracy is higher than that achieved by training on full image data. For the case of raw image data the AUC value increases from 0.911 to 0.920 whereas the AUC values for histogram equalisation and histogram specification improve to 0.945 and 0.955 respectively. These improvements can be attributed to the fact that PCA while discarding "less important" information essentially performs a noise filtering stage on the images. Applying both principle component analysis and histogram specification provides an optimal classification with a sensitivity of 92.31% and a specificity 92.41%.

Figure 3 on the left gives ROC curves of the average sensitivity and specificity from all 17 images based on a leave-one-out training scenario where, in turn, the network is trained on 16 images and then tested on the remaining image. Training raw colour data on the neural network gives an average AUC value of 0.927 which is slightly higher than the 0.911 achieved in the 10-2-5 testing scenario. Histogram equalisation again has a positive effect while histogram specification achieves the highest AUC of 0.951.

The ROC curves of the PCA networks on 16-1 training are shown on the right of Figure 3 and confirm that principle component analysis provides higher classification accuracy. The accuracy of the neural network improves greatly for all three different input types with histogram specification coupled with PCA providing the overall best classification performance with an AUC value of 0.973.

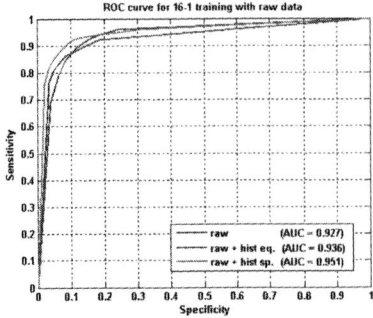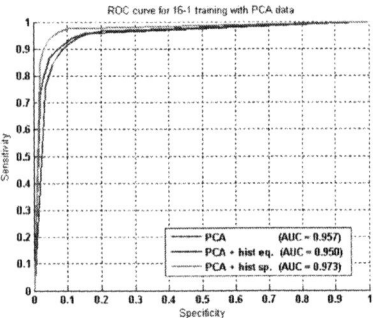

Fig. 3. ROC curves, based on 16-1 training, on raw data (left) and PCA data (right)

The optimum balanced sensitivity and specificity results for the best performing classifier give a sensitivity of 94.78% and a specificity of 94.29% which compare favourably with other results in the literature.

5   Conclusions

In this paper we have shown that a relatively "naïve" neural network can be successfully employed to detect of exudates in retinal images. A backpropagation neural network classifies central pixels of a sliding image window as being part of exudate or non-exudate regions. Histogram specification/equalisation is shown to be important to reach high classification rates while principal component analysis is applied for dimensionality reduction and reduced computational complexity. Despite the simplicity of the setup good classification performance is achieved providing a sensitivity of 94.78% with a specificity of 94.29% which compare well with other results in the literature.

References

1. Aiello, L., Callerano, J., Gardner, T., King, D., Blankenship, G., Ferris, F., Klein, R.: Diabetic retinopathy. Diabetes Care 21, 143–156 (1998)
2. Finlayson, G., Hordley, S., Schaefer, G., Tian, G.Y.: Illuminant and device invariant colour using histogram equalisation. Pattern Recognition 38, 179–190 (2005)
3. Gardner, G.G., Keating, D., Williamson, T.H., Elliott, A.T.: Automatic detection of diabetic retinopathy using an artificial neural network: a screening tool. British Journal of Ophthalmology 80(11), 940–944 (1996)
4. Goatman, K.A., Whitwam, A.D., Manivannan, A., Olson, J.A., Sharp, P.F.: Colour normalisation of retinal images. In: Medical Image Understanding and Analysis (2003)
5. Gonzales, R.C., Woods, R.E.: Digital Image Processing. Addison Wesley, Reading (1992)
6. Nabney, I.T.: Netlab: Algorithms for Pattern Recognition. Springer, Heidelberg (2002)

7. Osareh, A., Mirmehdi, M., Thomas, B., Markham, R.: Automated identification of diabetic retinal exudates in digital colour images. British Journal of Ophthalmology 87(10), 1220–1223 (2003)
8. Patton, N., Aslam, T.M., MacGillvray, T., Deary, I.J., Dhillon, B., Eikelboom, R.H., Yogesam, K., Constable, I.J.: Retinal image analysis: concepts, applications and potential. Progress in Retinal and Eye Research 25(1), 99–127 (2006)
9. Sinthanayothin, C., Boyce, J.F., Williamson, T.H., Cook, H.L., Mensah, E., Lal, S., Usher, D.: Automated detection of diabetic retinopathy on digital fundus images. Diabetic Medicine 19(2), 105–112 (2002)
10. Walter, T., Klein, J., Massin, P., Erginay, A.: A contribution of image processing to the diagnosis of diabetic retinopathy-detection of exudates in color fundus images of the human retina. IEEE Trans. Medical Imaging 21(10), 1236–1243 (2002)
11. Zweig, M.H., Campbell, G.: Receiver-operating characteristic (ROC) plots: a fundamental evaluation tool in clinical medicine. Clinical Chemistry 39(4), 561–577 (1993)

An Optimisation Approach to Palette Reordering for Indexed Image Compression

Gerald Schaefer[1] and Lars Nolle[2]

[1] School of Engineering and Applied Science
Aston University
Birmingham, U.K.
G.Schaefer@aston.ac.uk
[2] School of Computing and Informatics
Nottingham Trent University
Nottingham, U.K.
Lars.Nolle@ntu.ac.uk

Summary. Colour quantised images consist of two parts: a colour palette which represents all possible colours in the image, and an index image which stores which colour is to be used at each pixel position. This index image still contains a large amount of redundant information which can be exploited by applying lossless image compression algorithms. The achieved compression ratio however depends on the ordering of the colour entries in the palette. Hence, in order to achieve better compression, the colour entries need to be reordered so as to allow maximal redundancy exploitation by the specific compression algorithm to be used. In this paper we present a simulated annealing approach to reordering the palette for subsequent JPEG2000 or JPEG-LS lossless compression of the index information. We show that our algorithm allows for improved compression compared to previous reordering techniques.

Keywords: colour quantisation, colour palette, image compression, simulated annealing, optimisation analysis.

1 Introduction

Colour quantisation is a common image processing technique which can be used for displaying images on limited hardware such as mobile devices, for image compression, and for other applications such as image retrieval [15]. While true colour images typically use 24 bits per pixels (which results in an overall gamut of 2^{24} i.e. more than 16.7 million different colours), colour quantised images have only a limited palette of colours (usually between 8 and 256) that are used in the image. Various techniques for obtaining a good colour palette have been introduced in the literature (see e.g. [4, 2, 12]) and allow reasonable colour fidelity between original and colour quantised versions of an image.

A quantised image consist of two parts: a colour palette - obtained through application of a colour quantisation algorithm - which represents all possible colours in an image, and an index image which contains, for each pixel, a pointer to the palette identifying the colour to be used. Obviously, the index matrix still

E. Avineri et al. (Eds.): Applications of Soft Computing, ASC 52, pp. 178–185.
springerlink.com © Springer-Verlag Berlin Heidelberg 2009

contains a large amount of rendundant information as is the case with any kind of image. Applying a lossless compression algorithm such as JPEG2000 [6] or JPEG-LS [5] on the index data can thus be used to further compress palettised images. However, lossless image compression techniques are typically optimised for greyscale images and take into account certain image characteristics that allow good prediction of pixel values based on the intensities of neighbouring pixels. Consequently, smooth, continuous areas are compressed better than image regions that contain many intensity changes. In terms of the index information of colour quantised images, the information to be compressed does not directly translate to intensities but rather to the position of colours in the palette. It follows therefore that the actual compression performance depends not only on the image itself but also on the ordering of the colours in the palette as this in turn defines the actual "pixel" values of the index image [14]. Various specialised algorithms to this np-complete problem have been presented in the past in order to find an ideal ordering which allows good compressability of the index data. These methods can be divided into palette-based techniques [17, 3, 16] and index-based methods [8, 18, 13, 1].

In this paper we show how a generic black-box optimisation method, in particular a modified Simulated Annealing (SA) technique [10], can be used to generate a reordering of the palette entries for subsequent lossless compress of the index information. Experimental results prove that our optimisation based approach is capable of providing excellent compression capabilities, superior to those of most specialised algorithms.

The rest of the paper is organised as follows: in Section 2 we review some of the related work while in Section 3 we provide the background on optimisation based on Simulated Annealing and its variant that we employ. Section 4 explains our application of SA to the palette reordering problem. Section 5 gives experimental results while 6 concludes the paper.

2   Related Work

Various algorithms for palette reordering have been introduced in the past. In general, we can divide these into palette-based methods, and index-based approaches [14].

Palette-based algorithms perform the reording based solely on the colour palette itself. In its simplest form, originally proposed by Zaccarin and Liu [17], the palette colours are sorted in ascending order of their corresponding luminance values. Hadenfeldt and Sayood [3] applied a greedy algorithm which orders the colours so as to assign close indices to colours that are also close to each other in the 3-dimensional colour space. A similar approach was introduced by Spira and Malah [16] who sort the palette entries according to the distances between them in colour space by solving for a travelling salesman problem.

In contrast to the palette-based approaches, index-based methods also take the index information into account. The underlying idea of many of these methods is that colours that occur frequently close to each other in the image should be assigned close indices. Memon and Venkateswaran [8] reformulated the problem as one of linear predictive coding in which the objective is to minimise the zero-order entropy of prediction results, and a pairwise merging technique to obtain a near-optimal solution to the problem. Zeng et al. [18] suggested an algorithm that is based on a one-step look-ahead greedy approach but can be further improved as has been shown by Pinho and Neves [13]. Battiato et al. [1] pose the problem as that of finding the Hamiltonian path of maximum weight in a non-directed weighted graph and suggest a greedy algorithm to find the solution.

3   Simulated Annealing and SWASA

Simulated Annealing (SA) is an optimisation metaheuristic that was proposed by Kirkpatrick et al. [7] based on the work of Metropolis et al. [9]. SA is modelled after the physical process of cooling down a metal, which was heated up to a temperature near its melting point. If the subsequent cooling is carried out very slowly the particles tend to place themselves at positions where they reach a minimum energy state and hence form a more uniform crystalline structure.

The Simulated Annealing metaheuristic is based on the well-known hill-climbing algorithm. In hill climbing, a search starts from a randomly selected point from within the search space, which becomes the current solution. This current solution is then evaluated, i.e. the quality of the solution is measured in terms of an error value E. New trail solutions from the neighbourhood of the current solution are tested and if a candidate solution with a smaller error value is found this candidate solution becomes the current solution. Candidate solutions are usually chosen by moving a small step away from the current solution in a random directions, for example by adding small and equally distributed random numbers from the interval $[-s_{max}, s_{max}]$ to each component of the current solution vector. Here, s_{max} is called the 'maximum step width'. Obviously, values for s_{max} need to be chosen from the interval between 0 and the upper limit of the search space dimensions. A hill-climbing search is terminated either if the error value of the current solution is less that a predefined target value or the maximum number of iterations has been reached. The disadvantage of hill-climbing is that it tends to get stuck in the nearest minimum rather than finding the global minimum. This disadvantage is overcome in Simulated Annealing. If the difference in error ΔE between the current solution and the candidate solution is negative, i.e. the error of a trial solution is less than the error of the current solution, the trial solution is accepted as the current solution. But unlike hill-climbing, SA does not automatically reject a new candidate solution if ΔE is positive. Here it becomes the current solution with probability $p(T)$ which can be determined using

$$p(T) = e^{-\Delta E/T} \tag{1}$$

where T is referred to as 'temperature', an abstract control parameter for the cooling schedule.

For positive values of ΔE and a given temperature T, the probability function shown in Equation 1 has a defined upper limit of one and approaches zero for large positive values of ΔE. Therefore, in a computer simulation, the probability $p(T)$ has to be determined for each candidate solution and to be compared with an equally distributed random number drawn from the interval $[0, 1]$. If the probability $p(T)$ is greater than the random number the trial solution becomes the current solution, otherwise it is rejected.

The initial temperature has to be chosen in a way that leads to a high initial transition probability. The temperature is then reduced over the run towards zero according to a cooling schedule, which is usually chosen to be

$$T_{n+1} = \alpha T_n \qquad (2)$$

where T_n is the temperature at step n and α is the cooling coefficient (usually between 0.8 and 0.99).

During each step the temperature must be held constant for an appropriate number of iterations in order to allow the algorithm to settle into a 'thermal equilibrium' i.e. a balanced state. If the number of iterations is too small the algorithm is likely to converge to a local minimum. For both continuous parameter optimisation and discrete parameters with large search ranges, it is practically impossible to choose direct neighbours of the current solution as new trail solutions, because of the vast number of points in the search space. Therefore it is necessary to choose new candidates at some distance in a random direction of the current solution in order to travers in an acceptable time through the search space. This distance could either be a fixed step width s or it could be a random number with an upper limit s_{max}. In the first case, the neighbourhood would be defined as the surface of a hypersphere around the current solution, whereas in the latter the neighbourhood would be the volume of the hypersphere. New candidate solutions might be generated by adding small, equally distributed random numbers from the interval $[-s_{max}, s_{max}]$ to each component of the current solution vector. The maximum step width s_{max} is important to the success of SA. If s_{max} is chosen too small and the start point for a search run is too far away from the global optimum, the algorithm might not be able to get near that optimum before the algorithm 'freezes' i.e. the temperature becomes so small that $p(T)$ is virtually zero and the algorithm starts to perform only hill climbing.

If it did not pass the critical point before the algorithm changes to hillclimbing, it will get stuck in the nearest local optimum rather than finding the global one. If, on the other hand, the step width has been chosen to be too large and the peak of the optimum is very narrow, the algorithm might well get near the global optimum before the algorithm freezes, but never reaches the top because most of the steps are too large so that new candidate solutions fall off the peak. Therefore, there is always a trade-off between accuracy and robustness in selecting an appropriate maximum step width. If s_{max} is too small, SA has the potential to reach the peak of the 'frozen-in' optimum, but it cannot be guaranteed that this optimum is the global one. On the other hand, if s_{max} is too large, SA has the potential to get near the global optimum, but it might never reach the top of it.

One possible solution is to use small steps and to adjust the cooling schedule and therefore to increase the length of the Markov chains. This is not always possible in real-world optimisation problems with time constrains, i.e. where the number of fitness evaluations is limited. Step width adapting simulated annealing (SWASA) [10] overcomes the problems associated with constant values for s_{max} by using a scaling function [11] to adapt the maximum step width to the current iteration by

$$s_{max}(n) = \frac{2s_0}{1 + e^{\beta n / n_{max}}} \tag{3}$$

where $s_{max}(n)$ is the maximum step width at iteration n, s_0 is the initial maximum step width, n_{max} the maximum number of iterations and β is an adaptation constant.

4 Simulated Annealing for Colour Palette Reordering

In this paper we show that a generic black box optimisation algorithm can be employed to find a suitable palette entry reordering for subsequent lossless compression. The main advantage of black-box optimisation algorithms is that they do not require any domain specific knowledge yet are able to provide a near optimal solution. In particular, we apply the SWASA algorithm described in Section 3 to the colour palette reordering problem for indexed image compression. In contrast to the other techniques introduced in Section 2 which provide a specialised solution to the problem, the beauty of our approach is in its simplity as we do not need to encapsulate any domain specific information. All we have to define is an objective function to be optimised and a perturbation function which allows us to move in search space.

Since the aim is to obtain maximum compressibility, definition of the objective function is straightforward. We simply pass the index data image to a lossless compression algorithm and evaluate the resulting file size as the value of the objective function. We utilise two different compression techniques: JPEG2000 [6] and JPEG-LS [5]. Moving in search space is performed through pair-wise swap operations on the palette entries where the number of swap operations is related to the step width of the SWASA algorithm.

In our experiments we typically run the algorithm for 2000 iterations. Within each iteration we apply the pertubation function (i.e. perform swapping operations) and run the image compressor to calculate the corresponding value of the objective function for the newly generated solution. If this new solution is better than the previous one, it is adopted. If the new palette order offers worse compression it can still be adopted with a certain probability according to the current temperature T, as outlined in Section 3. At the end, the algorithm should have converged to an optimal or near-optimal palette reordering for the given image.

5   Experimental Results

In our experiments we use the *music* image (see Figure 1) which has been adopted frqeuently for evaluating palette reordering algorithms before [14]. The reason for this is that the image is fairly simple and that the optimal solution can be found through an exhaustive search as the palette only comprises 8 distinct colours (obviously even with only somewhat larger palettes exhaustive searches soon become infeasible due to the exponentially increasing number of possible reorderings) and is of size 111×111.

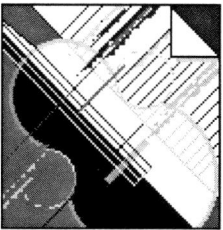

Fig. 1. *music* image

Leaving the colour palette in its original state and using JPEG2000 compression results in a representation that requires 1.708 bits per pixel (based on the file size of the JPEG2000 compressed index image and the uncompressed colour palette). In contrast, running an exhaustive search over all possible palette reorderings to retrieve the optimal configuration allows for at 1.374 bits per pixel is possible. Similarly, the optimum palette reordering for JPEG-LS compression offers a bitrate of 1.051, while performing compression based on the original palette gives 1.171 bits per pixel.

In Table 1 we provide full results of all the palette reordering methods that were introduced in Section 2. It is apparent that the colour-based approaches

Table 1. Compression results of various palette reordering methods on *music* image and JPEG2000/JPEG-LS compression

Palette reordering algorithm	bpp (JPEG2000)	bpp (JPEG-LS)
unordered	1.708	1.171
Zaccarin & Lui [17]	2.142	1.143
Hadenfeldt & Sayood [3]	1.794	1.287
Spira & Malah [16]	1.696	1.296
Memon & Venkateswaran [8]	**1.374**	**1.051**
Zeng *et al.* [18]	1.388	1.060
Pinho & Neves [13]	**1.374**	**1.051**
Battiato *et al.* [1]	1.428	1.071
proposed SWASA algorithm	**1.374**	**1.051**
optimal ordering (exhaustive search)	**1.374**	**1.051**

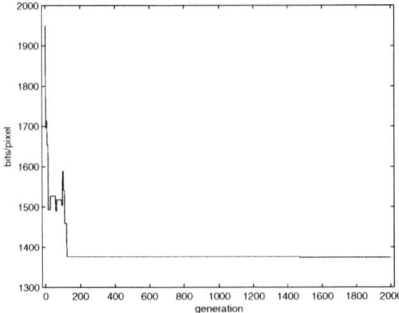

Fig. 2. Plot showing how the bitrate developes for a typical run of the algorithm on *music* image

perform rather poorly here. Indeed, for JPEG2000 coding only Spira and Malah's method introduces a slight improvement while the other two actually provide worse compression compared to the unordered palette. In the case of JPEG-LS Zaccarin and Lui's luminance based ordering afford slightly better compressability while here the other two techniques achieve worse bitrates.

In contrast, index-based methods clearly outperform palette-based approaches. Two of the methods, namely that by Memon and Venkateswaran and that by Pinho and Neves manage to find the optimal solution for both compression algorithms. The other two however, fail to do so despite the relatively small palette and image size.

Looking at the results achieved by our proposed generic optimisation approach, it can be seen that it too finds the optimal solution in both cases and indeed manages to do so without any domain specific knowledge. This proves that our proposed algorithm is not only able to provide a good solution for indexed image compression but that the obtained solutions are of the same quality as those of the best performing algorithms in the literature while furthermore outperforming several other specialised algorithms.

In Figure 2 we show a typical run of the algorithm by displaying the achieved bitrates during the 2000 iterations of the run. It can be seen that a near optimal solution (bpp = 1.376) is found very early, after about 120 generation whereas the global optimum of 1.374 is achieved after about 1500 generations.

6   Conclusions

Colour quantised images can be further compressed through the application of a lossless image compression algorithm on the index data. The resulting compression ratio however depends on the ordering of the colour entries in the palette. In this paper we have shown that a generic optimisation algorithm can be successfully employed to solve for the optimal palette reordering. Without the application of domain specific knowledge our proposed method is able to achieve compression equivalent to the best performing techniques in the literature while furthermore outperforming several other specialised algorithms.

References

1. Battiato, S., Gallo, G., Impoco, G., Stanco, F.: A color reindexing algorithm for lossless compression of digital images. In: Proc. IEEE Spring Conf. Computer Graphics, pp. 104–108 (2001)
2. Gervautz, M., Purgathofer, W.: A simple method for color quantization: Octree quantization. In: Glassner, A.S. (ed.) Graphics Gems, pp. 287–293 (1990)
3. Hadenfeldt, A.C., Sayood, K.: Compression of color-mapped images. IEEE Trans. Geoscience and Remote Sensing 32(3), 534–541 (1994)
4. Heckbert, P.S.: Color image quantization for frame buffer display. ACM Computer Graphics (ACM SIGGRAPH 1982 Proceedings) 16(3), 297–307 (1982)
5. ISO/IEC. Lossless and near-lossless compression of continuous-tone images - baseline. ISO/IEC International Standard 14495-1 (1999)
6. ISO/IEC. JPEG2000 image coding system: Core coding system. ISO/IEC International Standard 15444-1 (2002)
7. Kirkpatrick, S., Gelatt, C.D., Vecchi, M.P.: Optimization by simulated annealing. Science 220(4598), 671–680 (1983)
8. Memon, N.D., Venkateswaran, A.: On ordering color maps for lossless predictive coding. IEEE Trans. Image Processing 5(11), 1522–1527 (1996)
9. Metropolis, A., Rosenbluth, W., Rosenbluth, M.N., Teller, H., Teller, E.: Equation of state calculations by fast computing machines. Journal of Chemical Physics 21(6), 1087–1092 (1953)
10. Nolle, L.: On the effect of step width selection schemes on the performance of stochastic local search strategies. In: 18th European Simulation Multi-Conference, pp. 149–153 (2004)
11. Nolle, L., Goodyear, A., Hopgood, A.A., Picton, P.D., Braithwaite, N.: On step width adaptation in simulated annealing for continuous parameter optimisation. In: Reusch, B. (ed.) Fuzzy Days 2001. LNCS, vol. 2206, pp. 589–598. Springer, Heidelberg (2001)
12. Nolle, L., Schaefer, G.: Color map design through optimization. Engineering Optimization 39(3), 327–343 (2007)
13. Pinho, A.J., Neves, A.J.R.: A note on Zeng's technique for color reindexing of palette-based images. IEEE Signal Processing Letters 11(2), 232–234 (2004)
14. Pinho, A.J., Neves, A.J.R.: A survey on palette reordering methods for improving the compression of color-indexed images. IEEE Trans. Image Processing 13(11), 1411–1418 (2004)
15. Schaefer, G., Qiu, G., Finlayson, G.: Retrieval of palettised colour images. In: Wang, J., Yi, Z., Żurada, J.M., Lu, B.-L., Yin, H. (eds.) ISNN 2006. LNCS, vol. 3972, pp. 483–493. Springer, Heidelberg (2006)
16. Spira, A., Malah, D.: Improved lossless compression of color-mapped images by an approximate solution of the traveling salesman problem. In: IEEE Int. Conference on Acoustics, Speech and Signal Processing, vol. 3, pp. 1797–1800 (2001)
17. Zaccarin, A., Liu, B.: A novel approach for coding color quantized images. IEEE Trans. Image Processing 2(4), 442–453 (1993)
18. Zeng, W., Li, J., Lei, S.: An efficient color re-indexing scheme for palette-based compression. In: 7th IEEE Int. Conference on Image Processing, vol. III, pp. 476–479 (2000)

## Part VI

Production, Manufacturing, and Design

Weighted Tardiness Scheduling with Sequence-Dependent Setups: A Benchmark Problem for Soft Computing

Vincent A. Cicirello

Computer Science and Information Systems
The Richard Stockton College of New Jersey
Pomona, NJ 08240
cicirelv@stockton.edu

Summary. In this paper we present a set of benchmark instances and a benchmark instance generator for a single-machine scheduling problem known as the weighted tardiness scheduling problem with sequence-dependent setups. Furthermore, we argue that it is an ideal benchmark problem for soft computing in that it is computationally hard and does not lend itself well to exact solution procedures. Additionally, it has a number of important real world applications.

1 Introduction

In this paper we present a set of benchmark instances and a benchmark instance generator for a single-machine scheduling problem known as the weighted tardiness scheduling problem with sequence-dependent setups. Furthermore, we argue that it is an ideal benchmark problem for soft computing in that it is computationally hard and does not lend itself well to exact solution procedures. Additionally, it has a number of important real world applications.

The weighted tardiness problem is encountered in a number of real-world applications, including turbine component manufacturing [3], the packaging industry [1], among others [15]. It is a scheduling objective function that Morton and Pentico indicate to be very hard even if setups are independent of job sequence [15]. Exact optimal solutions can be found by branch-and-bound for instances of at most 100 jobs (e.g., [17]), but are considered impractical for instances that are larger than this [16]. For larger instances, heuristic or metaheuristic approaches are preferred (e.g., [9, 10, 16, 18, 11]).

What truly makes the problem a challenge problem is if you consider the case with sequence-dependent setups. *Setup time* refers to a length of time that must be spent preparing a machine prior to processing a job [15]. If all jobs are identical, or if the setup time only depends on the job that the machine is being setup for, but not on the previously processed job, then we can say that the setups are *sequence-independent*. If the setups are sequence-independent, then the problem can be transformed to essentially remove them from the problem (e.g., adding the setup times to the process times in some way). When setup time of a job depends on the job that is processed immediately before it on

E. Avineri et al. (Eds.): Applications of Soft Computing, ASC 52, pp. 189–198.
springerlink.com   © Springer-Verlag Berlin Heidelberg 2009

the machine then the setups are *sequence-dependent*. Allahverdi et al as well as Zhu and Wilhelm offer comprehensive reviews of these and other considerations pertaining to setup costs [2, 22].

Sequence-dependent setups commonly appear in real-world scheduling problems (e.g., [7, 1, 3, 14, 13]). Unfortunately, however, they are often ignored during the development of algorithms. The vast majority of work on sequencing problems assume that setups are sequence-independent, usually without acknowledging the possibility that they may be a factor in the problem. Sen and Bagchi discuss the significance of the challenge that sequence-dependent setups pose for exact solution procedures [21]. Specifically, they discuss how sequence-dependent setups induce a non-order-preserving property of the evaluation function. At the time of their writing, exact solution procedures such as A*, Branch-and-Bound algorithms, or GREC [21] for sequencing problems with sequence-dependent setups were limited to solving instances with no more than approximately 25-30 jobs, even for easier objective functions. Problem instances of larger size require turning to soft computing approaches.

Although there are exact approaches for optimizing weighted tardiness when setups are independent of sequence (or non-existent), all current approaches for the problem when setups are sequence-dependent are either heuristic or metaheuristic. For example, there are some dispatch scheduling heuristics for the sequence-dependent setup version of the weighted tardiness problem such as ATCS [12] as well as the heuristic of [20]. Both of these are rather ad hoc modifications of the well-known R&M dispatch policy [19] for the setup-free version of the problem. Additionally, there have been several recent metaheuristics for the problem. Lee et al suggested a local hill climbing algorithm to apply to the solution given by their ATCS dispatch policy [12]. Cicirello and Smith developed a value-biased stochastic sampling algorithm to expand the search around ATCS; and also benchmarked their approach with several other heuristic search algorithms [8]. Most recently, a permutation-based genetic algorithm using the Non-Wrapping Order Crossover operator [5] and a simulated annealing algorithm [6] have both improved upon a number of the best known solutions to several benchmark instances.

2   Problem Formulation

The weighted tardiness scheduling problem with sequence-dependent setups can be defined as follows. The problem instance consists of a set of jobs $J = \{j_0, j_1, \ldots, j_N\}$. Each of the jobs j has a weight w_j, duedate d_j, and process time p_j. Furthermore, $s_{i,j}$ is defined as the amount of setup time required immediately prior to the start of processing job j if it follows job i on the machine. It is not necessarily the case that $s_{i,j} = s_{j,i}$. The 0-th "job" is the start of the problem ($p_0 = 0$, $d_0 = 0$, $s_{i,0} = 0$, $w_0 = 0$). Its purpose is for the specification of the setup time of each of the jobs if sequenced first.

The weighted tardiness objective is to sequence the set of jobs J on a machine to minimize:

$$T = \sum_{j \in J} w_j T_j = \sum_{j \in J} w_j \max(c_j - d_j, 0), \qquad (1)$$

where T_j is the tardiness of job j; and c_j, d_j are the completion time and duedate of job j. The completion time of a job is equal to the sum of the process times and setup times of all jobs that come before it in the sequence plus the setup time and process time of the job itself. Specifically, let $\pi(j)$ be the position in the sequence of job j. Define c_j as:

$$c_j = \sum_{i,k \in J, \pi(i) <= \pi(j), \pi(i) = \pi(k)+1} p_i + s_{k,i}. \qquad (2)$$

3   Weighted Tardiness Problem Instance Generator

Previously, the author implemented a problem instance generator for the weighted tardiness scheduling problem with sequence-dependent setups [4]. This instance generator is an implementation of a procedure described by Lee et al and used in the analysis of Lee et al's dispatch scheduling policy ATCS [12]. Cicirello's instance generator has since been refined and reimplemented in Java and is available on the web (http://loki.stockton.edu/~cicirelv/benchmarks/).

Each problem instance is characterized by three parameters: the due-date tightness factor τ; the due-date range factor R; and the setup time severity factor η. These parameters are defined as follows:

$$\tau = 1 - \frac{\bar{d}}{C_{\max}} \qquad (3)$$

$$R = \frac{d_{\max} - d_{\min}}{C_{\max}} \qquad (4)$$

$$\eta = \frac{\bar{s}}{\bar{p}} \qquad (5)$$

where $\bar{d}$, $\bar{p}$, and $\bar{s}$ are the average duedate, average process time, and average setup time, $d_{\max}$, $d_{\min}$ are the maximum and minimum duedates, and $C_{\max}$ is the makespan (or completion time of the last job). Given that the makespan depends on the optimal sequence and the $s_{i,j}$, the estimator suggested by Lee et al is used: $\tilde{C}_{\max} = n(\bar{p} + \beta \bar{s})$ where n is the number of jobs in the problem instance.

Lee et al provide experimental data for setting the value of β for 4 different size problems (20, 40, 60, and 80 job instances). Cicirello's original implementation of the instance generator was restricted to generating instances of those 4 sizes. One of the refinements of the new Java implementation is fitting a curve to Lee et al's reported data to extrapolate appropriate values of β for other size problem instances. Specifically, a small 3 node feedforward neural net (2 hidden sigmoid nodes, 1 output sigmoid) with the number of jobs, N, as input was fitted to the data through a least-squares fit. The result is the following definition of β as a function of the number of jobs:

$$\beta(N) = \frac{1}{1 + e^{(1.0949132 - 1971.6253 \cdot A(N) - 8.1243637 \cdot B(N))}}, \tag{6}$$

with

$$A(N) = \frac{1}{1 + e^{(7.168150953 + 0.040112027 \cdot N)}} \tag{7}$$

and

$$B(N) = \frac{1}{1 + e^{(-10.58867025 + 2.400027877 \cdot N)}}. \tag{8}$$

The processing times of the jobs of the instances produced by the generator are uniformly distributed over the interval $[50, 150]$, with $\bar{p} = 100$. The mean setup time $\bar{s}$ is then determined from η and the setup times are uniformly distributed in the interval $[0, 2\bar{s}]$. The duedate of a job is uniformly distributed over $[\bar{d}(1 - R), \bar{d}]$ with probability τ and uniformly distributed over $[\bar{d}, \bar{d} + (C_{\max} - \bar{d})R]$ with probability $1 - \tau$. The weights of the jobs are distributed uniformly over $[0, 10]$.

4   Weighted Tardiness Benchmark Instances

In addition to the instance generator, the set of benchmark problem instances used by [4, 8, 5, 6] among others, are available on the web (http://loki.stockton.edu/~cicirelv/benchmarks/). This benchmark set includes 120 problem instances with 60 jobs each. The problem set is characterized by the following parameter values: $\tau = \{0.3, 0.6, 0.9\}$; $R = \{0.25, 0.75\}$; and $\eta = \{0.25, 0.75\}$. For each of the twelve combinations of parameter values, there are 10 problem instances. Generally speaking, these 12 problem sets cover a spectrum from loosely to tightly constrained problem instances.

Table 1. Current best known solutions for instances with: (a) Loose Duedates, Narrow Duedate Range, Mild Setups; and (b) Loose Duedates, Narrow Duedate Range, Severe Setups

(a)				(b)		
Instance	Best	First Found By		Instance	Best	First Found By
1	790	SA-H		11	5088	SA-H
2	5824	SA-H		12	0	LDS
3	1936	GA		13	6147	VBSS-HC
4	6840	SA-H		14	3761	SA-R
5	5017	SA-R		15	2039	SA-R
6	7824	SA-H		16	5559	SA-R
7	3933	SA-H		17	387	SA-R
8	298	SA-R		18	1918	SA-R
9	7059	SA-H		19	239	SA-H
10	2125	SA-H		20	3805	SA-R

Table 2. Current best known solutions for instances with: (a) Loose Duedates, Wide Duedate Range, Mild Setups; and (b) Loose Duedates, Wide Duedate Range, Severe Setups

(a)				(b)		
Instance	Best	First Found By		Instance	Best	First Found By
21	0	LDS		31	0	LDS
22	0	LDS		32	0	LDS
23	0	LDS		33	0	LDS
24	1092	SA-H		34	0	LDS
25	0	SA		35	0	LDS
26	0	LDS		36	0	LDS
27	57	SA-R		37	1008	SA-R
28	0	GA		38	0	LDS
29	0	LDS		39	0	LDS
30	215	SA-R		40	0	LDS

File Format

Each benchmark instance is stored in a separate file according to the following file format:

```
Problem Instance: <instance number>
Problem Size: <number of jobs>
Begin Generator Parameters
Tau: <tau>
R: <R>
Eta: <eta>
P_bar: <average process time>
P_MIN: <minimum process time>
P_MAX: <maximum process time>
S_bar: <average setup time>
MAX_WEIGHT: <maximum weight value>
C_max: <makespan estimate>
D_bar: <average duedate>
End Generator Parameters
Begin Problem Specification
Process Times:
<process time for job 0>
...
<process time for job n-1>
Weights:
<weight for job 0>
...
<weight for job n-1>
Duedates:
<duedate for job 0>
```

Table 3. Current best known solutions for instances with: (a) Moderate Duedates, Narrow Duedate Range, Mild Setups; and (b) Moderate Duedates, Narrow Duedate Range, Severe Setups

(a)				(b)		
Instance	Best	First Found By		Instance	Best	First Found By
41	71242	SA-H		51	54707	SA-H
42	59493	SA-H		52	100793	SA-H
43	147737	SA-H		53	94394	SA-R
44	36265	SA-H		54	123558	VBSS
45	59696	SA-R		55	72420	SA-R
46	36175	SA-R		56	80258	SA-H
47	74389	SA-H		57	68535	SA-H
48	65129	SA-R		58	46978	SA-R
49	79656	SA-R		59	56181	SA-H
50	32777	SA-R		60	68395	SA-R

```
. . .
<duedate for job n-1>
Setup Times:
<i> <j> <setup for job j if after i>
// i=-1 indicates setup if j is first
End Problem Specification
```

Current Best Known Solutions.

Tables 1(a), 1(b), 2(a), 2(b), 3(a), 3(b), 4(a), 4(b), 5(a), 5(b), 6(a), and 6(b) list the current best known solutions to the benchmark instances. Specifically, each table lists the instance number, the current best known solution (for the weighted tardiness objective), and the algorithm that first found that solution. Note that other algorithms may also have found that solution. We simply list here the first one that did. Algorithms are abbreviated as follows:

- SA-H: A recent simulated annealing algorithm [6].
- SA-R: A second version of that recent simulated annealing algorithm [6].
- GA: A permutation-based genetic algorithm using the Non-Wrapping Order Crossover operator and an insertion mutator [5].
- VBSS-HC: A multistart hill climber seeded with starting configurations generated by Value Biased Stochastic Sampling [8].
- VBSS: Value Biased Stochastic Sampling [8].
- LDS: Limited Discrepancy Search truncated after exhausting all search trajectories with 2 or less discrepancies from the ATCS heuristic solution [8]. It was used in the original empirical evaluation of VBSS and VBSS-HC.
- SA: A simulated annealing algorithm also used in the original empirical evaluation of VBSS and VBSS-HC [8].

Table 4. Current best known solutions for instances with: (a) Moderate Duedates, Wide Duedate Range, Mild Setups; and (b) Moderate Duedates, Wide Duedate Range, Severe Setups

(a)			(b)		
Instance	Best	First Found By	Instance	Best	First Found By
61	76769	SA-R	71	155036	SA-R
62	44781	SA-H	72	49886	SA-H
63	76059	SA-H	73	30259	SA-H
64	93079	SA-H	74	32083	SA-R
65	127713	SA-R	75	21602	SA-R
66	59717	SA-H	76	57593	SA-H
67	29394	SA-R	77	35380	SA-H
68	22653	SA-R	78	21443	SA-H
69	71534	SA-H	79	121434	SA-H
70	76140	SA-R	80	20221	SA-H

Table 5. Current best known solutions for instances with: (a) Tight Duedates, Narrow Duedate Range, Mild Setups; and (b) Tight Duedates, Narrow Duedate Range, Severe Setups

(a)			(b)		
Instance	Best	First Found By	Instance	Best	First Found By
81	385918	SA-H	91	344428	SA-R
82	410550	SA-H	92	363388	SA-R
83	459939	SA-R	93	410462	VBSS
84	330186	SA-R	94	334180	SA-H
85	557831	SA-R	95	524463	SA-R
86	364474	SA-R	96	464403	LDS
87	400264	SA-R	97	418995	SA-H
88	434176	SA-R	98	532519	VBSS
89	411810	SA-H	99	374607	SA-R
90	403623	SA-R	100	441888	VBSS-HC

The problem instances and the best known solutions to them are organized according to the 12 classes of instances as follows:

- Table 1(a): Loose Duedates, Narrow Duedate Range, Mild Setups
- Table 1(b): Loose Duedates, Narrow Duedate Range, Severe Setups
- Table 2(a): Loose Duedates, Wide Duedate Range, Mild Setups
- Table 2(b): Loose Duedates, Wide Duedate Range, Severe Setups
- Table 3(a): Moderate Duedates, Narrow Duedate Range, Mild Setups
- Table 3(b): Moderate Duedates, Narrow Duedate Range, Severe Setups
- Table 4(a): Moderate Duedates, Wide Duedate Range, Mild Setups
- Table 4(b): Moderate Duedates, Wide Duedate Range, Severe Setups

Table 6. Current best known solutions for instances with: (a) Tight Duedates, Wide Duedate Range, Mild Setups; and (b) Tight Duedates, Wide Duedate Range, Severe Setups

(a)				(b)		
Instance	Best	First Found By		Instance	Best	First Found By
101	353575	SA-R		111	348796	SA-H
102	495094	SA-H		112	375952	SA-H
103	380170	VBSS		113	261795	SA-H
104	358738	SA-R		114	471422	SA-H
105	450806	SA-R		115	460225	VBSS
106	457284	SA-H		116	537593	SA-H
107	353564	SA-H		117	507188	SA-H
108	462675	SA-R		118	357575	LDS
109	413918	SA-H		119	581119	SA-H
110	419014	SA-R		120	399700	VBSS-HC

- Table 5(a): Tight Duedates, Narrow Duedate Range, Mild Setups
- Table 5(b): Tight Duedates, Narrow Duedate Range, Severe Setups
- Table 6(a): Tight Duedates, Wide Duedate Range, Mild Setups
- Table 6(b): Tight Duedates, Wide Duedate Range, Severe Setups

In this set of benchmark instances, loose duedates, moderate duedates, and tight duedates refer to values of the duedate tightness factor τ of 0.3, 0.6, and 0.9, respectively. Narrow duedate range and wide duedate range refer to values of the duedate range factor of 0.25 and 0.75, respectively. Mild setups and severe setups refer to values of the setup time severity factor η of 0.25 and 0.75, respectively.

5 Conclusions

In this paper, we presented a set of benchmark instances and a problem instance generator for a computationally hard scheduling problem known as the weighted tardiness scheduling problem with sequence-dependent setups. Two characteristics of the problem make it ideal for benchmarking soft computing algorithms. First, the sequencing jobs on a single machine to optimize this objective function is NP-Hard even in the setup-free version. Exact approaches are currently limited to instances with no more than 100 jobs; and are infeasible beyond that size. Second, sequence-dependent setups greatly magnify the problem difficulty. It has been shown by others that sequence-dependent setups, even for easier objective functions than weighted tardiness, induce a non-order-preserving property for exact approaches, largely limiting them to solving instances smaller than approximately 30 jobs. Even moderately sized instances require metaheuristics and other soft computing approaches.

References

1. Adler, L., Fraiman, N.M., Kobacker, E., Pinedo, M., Plotnitcoff, J.C., Wu, T.P.: BPSS: a scheduling system for the packaging industry. Operations Research 41, 641–648 (1993)
2. Allahverdi, A., Guptab, J.N.D., Aldowaisan, T.: A review of scheduling research involving setup considerations. Omega: International Journal of Management Science 27, 219–239 (1999)
3. Chiang, W.Y., Fox, M.S., Ow, P.S.: Factory model and test data descriptions: OPIS experiments. Technical Report CMU-RI-TR-90-05, The Robotics Institute, Carnegie Mellon University (March 1990)
4. Cicirello, V.A.: Boosting Stochastic Problem Solvers Through Online Self-Analysis of Performance. PhD thesis, The Robotics Institute, School of Computer Science, Carnegie Mellon University, Pittsburgh, PA, 21, Also available as technical report CMU-RI-TR-03-27 (July 2003)
5. Cicirello, V.A.: Non-wrapping order crossover: An order preserving crossover operator that respects absolute position. In: Keijzer, M., et al. (eds.) Proceedings of the Genetic and Evolutionary Computation Conference (GECCO 2006), vol. 2, pp. 1125–1131. ACM Press, New York (2006)
6. Cicirello, V.A.: On the design of an adaptive simulated annealing algorithm. In: Proceedings of the CP 2007 First International Workshop on Autonomous Search, September 23 (2007)
7. Cicirello, V.A., Smith, S.F.: Wasp-like agents for distributed factory coordination. Journal of Autonomous Agents and Multi-Agent Systems 8(3), 237–266 (2004)
8. Cicirello, V.A., Smith, S.F.: Enhancing stochastic search performance by value-biased randomization of heuristics. Journal of Heuristics 11(1), 5–34 (2005)
9. Cicirello, V.A., Smith, S.F.: The max k-armed bandit: A new model of exploration applied to search heuristic selection. In: Veloso, M.M., Kambhampati, S. (eds.) The Proceedings of the Twentieth National Conference on Artificial Intelligence, vol. 3, pp. 1355–1361. AAAI Press, Menlo Park (2005)
10. Congram, R.K., Potts, C.N., van de Velde, S.L.: An iterated dynasearch algorithm for the single-machine total weighted tardiness scheduling problem. INFORMS Journal on Computing 14(1), 52–67 (2002)
11. Crauwels, H.A.J., Potts, C.N., Van Wassenhove, L.N.: Local search heuristics for the single machine total weighted tardiness scheduling problem. INFORMS Journal on Computing 10(3), 341–350 (1998)
12. Lee, Y.H., Bhaskaran, K., Pinedo, M.: A heuristic to minimize the total weighted tardiness with sequence-dependent setups. IIE Transactions 29, 45–52 (1997)
13. Morley, D.: Painting trucks at general motors: The effectiveness of a complexity-based approach. In: Embracing Complexity: Exploring the Application of Complex Adaptive Systems to Business, pp. 53–58. The Ernst and Young Center for Business Innovation (1996)
14. Morley, D., Schelberg, C.: An analysis of a plant-specific dynamic scheduler. In: Proceedings of the NSF Workshop on Intelligent, Dynamic Scheduling for Manufacturing Systems, pp. 115–122 (June 1993)
15. Morton, T.E., Pentico, D.W.: Heuristic Scheduling Systems: With Applications to Production Systems and Project Management. John Wiley and Sons, Chichester (1993)
16. Narayan, V., Morton, T., Ramnath, P.: X-Dispatch methods for weighted tardiness job shops. GSIA Working Paper 1994-14. Carnegie Mellon University, Pittsburgh (July 1994)

17. Potts, C.N., van Wassenhove, L.N.: A branch and bound algorithm for the total weighted tardiness problem. Operations Research 33(2), 363–377 (1985)
18. Potts, C.N., Van Wassenhove, L.N.: Single machine tardiness sequencing heuristics. IIE Transactions 23(4), 346–354 (1991)
19. Rachamadugu, R.V., Morton, T.E.: Myopic heuristics for the single machine weighted tardiness problem. Working Paper 30-82-83, GSIA. Carnegie Mellon University, Pittsburgh, PA (1982)
20. Raman, N., Rachamadugu, R.V., Talbot, F.B.: Real time scheduling of an automated manufacturing center. European Journal of Operational Research 40, 222–242 (1989)
21. Sen, A.K., Bagchi, A.: Graph search methods for non-order-preserving evaluation functions: Applications to job sequencing problems. Artificial Intelligence 86(1), 43–73 (1996)
22. Zhu, X., Wilhelm, W.E.: Scheduling and lot sizing with sequence-dependent setup: A literature review. IIE Transactions 38(11), 987–1007 (2006)

An Exploration of Genetic Process Mining

Chris Turner and Ashutosh Tiwari

School of Applied Sciences, Cranfield University,
Cranfield, Bedfordshire, UK
Tel.: +44 1234 758250; Fax: +44 1234 754604
c.j.turner@cranfield.ac.uk,
a.tiwari@cranfield.ac.uk

Abstract. In this paper the practice of business process mining using genetic algorithms is described. This paper aims to ascertain the optimum values for two fitness function parameters within a process mining genetic algorithm; the first parameter reduces the likelihood of process models with extra behaviour being selected and similarly the second parameter restricts the selection of models containing duplicate tasks. The experiments conducted in this research also include the use of a decaying rate for the mutation operator in order to promote greater accuracy in the mined process models. Details are provided on the fitness and mutation algorithms employed. The paper concludes that the optimum setting of the fitness function parameters will in fact vary depending on the constructs found in each process model. This paper finds that a higher value for one of the fitness function parameters allows for simple process constructs to be mined with greater accuracy. The use of a decaying rate of mutation is also found to be beneficial in the correct mining of simple processes.

Keywords: Business process mining, Genetic algorithms, Mutation rate, Chromosome fitness measures, Business process management.

1 Introduction

The practice of business process mining derives from the field of data mining. Data mining refers to the extraction of knowledge from large data sets though identification of patterns within the data. Data mining practice has been developed and adapted to create the business process mining techniques that are now being used to mine data logs containing process execution data to reconstruct actual business processes. These data logs are typically hosted within Business Process Management (BPM) and workflow systems, though they may also be accessible through other process related systems installed within a company.

The need for companies to learn more about how their processes operate in the real world is a major driver behind the development and increasing use of process mining techniques. Currently many approaches to process mining make use of heuristic algorithms ('rules of thumb' based on assumptions about business process patterns). Process mining is uniquely challenging as there is often just one correct representation of a process which must be reconstructed from process execution traces. Often the data logs containing these execution traces are noisy and incomplete. Genetic Algorithms (GA) are useful in such a situation as they are more resilient to noisy data and are also

E. Avineri et al. (Eds.): Applications of Soft Computing, ASC 52, pp. 199–208.

able to mine more complex process constructs such as loops and cope with the presence of duplicate tasks, among other common process mining challenges [2]. This paper focuses on the use of genetic algorithm techniques for process mining, and investigates the effect of amending the mutation and fitness functions within a process mining genetic algorithm.

2   A Genetic Process Mining Algorithm

The key stages of the process mining algorithm are detailed in Alves de Medeiros et al. [1] and are summarised as: 1) Read the Event Log, 2) Build the Initial Population, 3) Calculate Individuals' Fitness, 4) Stop and Return the Fittest Individuals and 5) Create Next Population. Of particular interest to this paper are the algorithmic functions put forward by Alves de Medeiros [1] to address the aspects of mutation and fitness.

$$F_{itness}(L,CM,CM[]) = F(L,CM,CM[]) - \gamma * PF \ folding(CM,CM[])$$

Where

$$F(L,CM,CM[]) = PF \ complete(L,CM) - \kappa * PF \ precise(L,CM,CM[]) \qquad (1)$$

In the algorithm shown in equation 1, L is an event log and CM a process model (a process model represented by a causal matrix). CM[] represents a generation of process models. This equation gives a fitness value for each individual process model within a generation. The κ parameter acts as a weighting for process models that allow for more transitions between process activities than is recorded in the event log (extra behaviour). This reduces the likelihood of such process models being selected for either crossover or elite status. The γ parameter acts as a weighting for process models that allow for the presence of duplicate tasks (the same task appearing multiple times within a process often indicates an incorrect process model). The genetic algorithm is only able to differentiate between duplicate tasks by examining the local context of each duplicate; the input and output activities for a duplicate activity. When duplicate activities share output and/or input tasks this differentiation is not possible so this trait is punished by the fitness algorithm [1] (for a more detailed explanation of the fitness function outlined by equation 1 see [1]).

Of particular interest in this paper are the κ and γ parameters (equation 1). As mentioned before the κ parameter is used in the weighting of the extra behaviour punishment measure, and the γ parameter is used in the weighting of the punishment for duplicate tasks that share input and/or output tasks. The punishment function is detailed in [1].

The genetic algorithm outlined by Alves de Medeiros [1] also includes a mutation function which can change, at random, different sections of a process model. Three different actions are performed by the mutation function on process models: 1) An activity is removed 2) An activity is added 3) Activities are re-organised. The choice of which mutation action is selected is random. A mutation rate is used to control the amount of mutation that may occur at each generation. This rate is constant throughout the running of the genetic algorithm. In order to test the overall correctness of a process model produced by running the genetic process mining algorithm four measures have been produced by Alves de Medeiros [1]; Two to test the behaviour of the end process model and two to examine the structure of that model. These measures are used in this paper's experiment results. Throughout the experiments detailed in this paper the duplicate task compliant version of the GA provided by [1] has been used.

3 An Experimental Evaluation

This paper studies the effect of applying a decaying rate of mutation to the mutation function employed by Alves de Medeiros [1]. Authors on evolutionary techniques, such as Cox [3], point to the use of a decaying rate of mutation in genetic algorithms. It is claimed that such a function may help the genetic algorithm to converge on a solution sooner by gradually reducing genetic diversity, from a relatively high base, over the generations.

$$Pm = Pm \times min \left(\frac{l}{g^c}, 1 \right) \tag{2}$$

In equation 2 above, the non-uniform decay algorithm of Cox [3], is shown. The notation Pm represents the probability that an individual will mutate. The notation l represents the switchover limit, the mutation rate stays at its initial value until $g^c = l$ (l is set to 1 in this paper). The notation g^c is a generation counter. The proposed algorithm for mutation is written as pseudo code –

```
MutationR = MutationR * min(1.0 / GenerationCount)
If MutationR < newRandomD Then
      If  MutationR > new(RandomD) then
            Case DeleteActivity if RandomD > 0.2
            Case  InsertActivity > 0.3
            Case RearrangeElements > 0.5
      End select
End if
Return Mutation choice
```

Fig. 1. Implementation of Mutation Algorithm

In Fig. 1 above the notation MutationR refers to the mutation rate calculation. The notation new(RandomD) indicates a function that returns RandomD as a new random number between 0.0 and 1.0. A case statement exists to randomly select the method of mutation from the three choices. The first two choices in the case statement will delete and insert an extra activity, respectively, if the random number is greater than their selection value. The third choice, if selected, will rearrange the elements of a process model, swapping the positions of a number of tasks.

The pseudo code from the second line describes the current operation of the mutation function outlined in Alves de Medeiros [1]. This paper also experiments with the κ and γ parameters in the fitness measure employed by Alves de Medeiros [1]. As mentioned earlier, in the Related Work section, the κ parameter is used in the weighting of the extra behaviour punishment measure, and the γ parameter is used in the weighting of the punishment for duplicate tasks that share input and/or output tasks. It is noted by [1] that optimal values have not been obtained for these two parameters. Optimal values for these punishment parameters are required as the presence of both duplicates and extra behaviour (behaviour that is not recorded in the event log) is detrimental to the overall correctness of a process model. The summary of this fitness measure is presented in equation 1. It is intended that a range of values are used to represent both κ and γ parameters in order to ascertain their optimal setting. Both parameters are set at 0.025 in the work of Alves de Medeiros [1]. The open source ProM Process mining framework [5] will be used to perform the experiments as it contains the process mining genetic algorithm set out by Alves de Medeiros [1]. The process model measures outlined in section 3 of this paper will be used to quantify the effects of the experiments to be performed.

4 Design of Experiments

The experiments with the values used to represent both κ and γ parameters will focus on testing values between 0.01 and 0.03. The current setting for both values is 0.025 as demonstrated in Alves de Medeiros [1], and therefore represents the as-is state. Table 1 below shows the values that will be used in the first run of the experiments for parameter κ, with parameter γ remaining at 0.025 (Table 1, tests 1-6). The values for the two parameters have been chosen to represent a wide range of weighting values in order to determine the optimum setting (the value ranges were selected due to their performance in preliminary tests). Tests 7 – 12 show parameter γ varying with

Table 1. Tests for parameter κ & γ

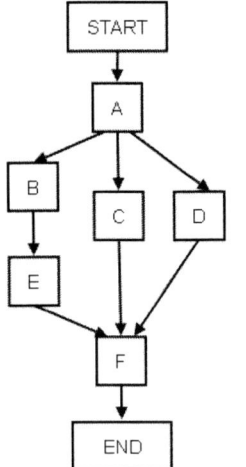

Fig. 2. Template for process data 1

	Parameter κ	Parameter γ
Test 1	0.018	0.025
Test 2	0.020 (As-Is)	''
Test 3	0.022	''
Test 4	0.024	''
Test 5	0.025	''
Test 6	0.027	''
Test 7	0.025	0.023
Test 8	''	0.024
Test 9	''	0.025 (As-Is)
Test 10	''	0.027
Test 11	''	0.029
Test 12	''	0.03
Test 13	0.018	0.023
Test 14	0.02	0.024
Test 15	0.022	0.025 (As-Is)
Test 16	0.024	0.027
Test 17	0.025	0.029
Test 18	0.027	0.03

parameter κ remaining at 0.025. Tests 13 – 18 show both parameters set varying values. The tests shown in Table 1 will be repeated for each of the five sets of process data in Table 2. Each process data set comprises a process event log and a completed process model template (shown in Fig. 2.), detailing the complete process.

These models represent a selection of process conditions containing varying levels of sequential, parallel and looping process structure. The process data represents a carefully selected set of tests that were used to test the original genetic process mining algorithm set out by Alves de Medeiros [1].

Table 2. Test processes to be used in the experiments (the process template for each process data set is shown in the appendices)

	Sequences	Parallelism	Loops
Process Data 1	Low	Low	Low
Process Data 2	Low	Low	Medium
Process Data 3	Medium	Medium	Low
Process Data 4	High	Low	Low
Process Data 5	Low	High	Low

Similarly a range of values will be used to test the decaying rate mutation enhancement outlined in section 3 of this paper. Table 3, sets out the range of values to be used in testing this enhancement. The initial run of the experiment will use the mutation rates set out in Table 3 without decay to establish an as-is state (which includes the current setting of 0.2 as used by Alves de Medeiros [1]. This experiment will be repeated with each of the five data sets listed in Table 2. The second run of the experiment will involve the use of values set out above in Table 3 with the decay enhancement. Again this experiment will be repeated with each of the five data sets listed in Table 2.

The choice of the values for the mutation experiment have been influenced by Negnevitsky [4], who suggests the mutation probability for genetic algorithms is usually between 0.001 and 0.01.

Cox [3] also suggests the following formula for determining a suitable rate of mutation, shown below in equation 3.

$$m_r = \max\left(0.01, \frac{1}{N} \right) \tag{3}$$

In equation 3 above m_r represents the probability of mutation or mutation rate. The notation N represents the population size. In most of the experiments carried out by Alves de Medeiros [1] the population size was 100, giving a mutation rate of 0.01 by the formula.

However Alves de Medeiros [1] recommends a mutation rate of 0.02. It is known from the work of this author that in building the initial population heuristics are used to determine causality relations between process tasks; though non local causality relations cannot be inferred in this way. Thus the non local causality relations (relations that cannot be inferred by just examining the direct predecessors and successors of a task) are created by the mutation operator. In the experiments a range of high and

Table 3. Tests for the rate of mutation

	Mutation rate starting value		Mutation rate starting value
Test 1	0.1	Test 8	0.25
Test 2	0.15	Test 9	0.27
Test 3	0.18	Test 10	0.28
Test 4	0.2	Test 11	0.3
Test 5	0.21	Test 12	0.35
Test 6	0.22	Test 13	0.4
Test 7	0.24	Test 14	0.5

low values for the mutation operator (relatively high given the previous discussion) will be used. The following settings were used in the running of the genetic process mining algorithm (within the ProM process mining framework [5]) : *Population size 100, Maximum generations 1000, Elitism rate 0.02, Crossover rate 0.8.* The fitness type was set to 'ExtraBehaviour'. The duplicate Tasks version of the genetic process mining algorithm was used in the completion of the experiments in this paper. Preliminary test were undertaken to establish a range of values to be used in the experiments. This involved the use of process data 1. When all of the four measures (two for process behaviour and two for structure) used in the experiments are equal to 1 the mined process model has perfect correspondence to the template model (see section 3.2 for descriptions of these measures).

5 Experimental Results

5.1 Results of Fitness Function Experiments

It was found that the use of different values in the setting of the κ and γ variables of the fitness function was beneficial in the mining of process data 1, 2 and 3. The only correct model that could be mined by the use of the standard setting of 0.25 (for both parameters) was process data 3. Fig. 3 below shows the results for the use of a range of values for both parameters with process data 1. The use of a slightly higher value, than the standard setting, for γ with κ remaining at 0.25 (or slightly lower) allowed for the correct mining of the process model (shown in Fig. 3 below). From tests with just the γ value changing a level between 0.026 and 0.03 was required for the correct mining of process data 1 (Fig. 4). Fig. 5 shows that just varying κ does not allow for the correct mining of process data 1.

The results for process data 2 again showed that a slightly higher value for γ could allow for the correct mining of the process model. Process data 3 could be mined correctly with almost every combination of values. The process models resulting for process data 4 and 5 (results for process data 5 are shown in Fig. 6) were the most difficult to mine, being highly parallel and sequential in nature. Neither process model could be mined correctly; though the mining of process data 4 did benefit from a slightly higher than recommended value for γ and lower than recommended value for κ.

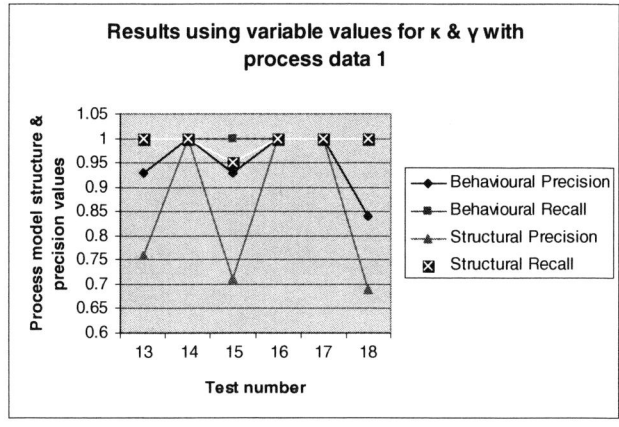

Fig. 3. Results using variable values for κ & γ (process data 1)

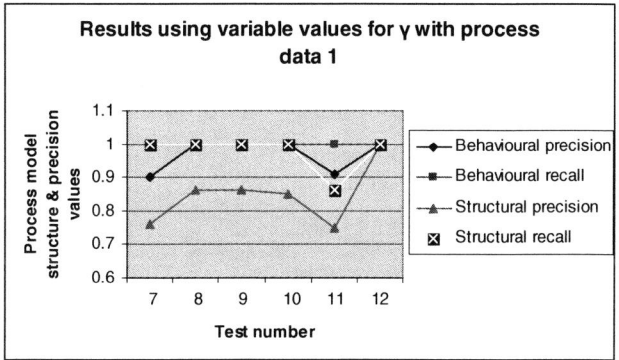

Fig. 4. Results using variable values for γ (process data 1)

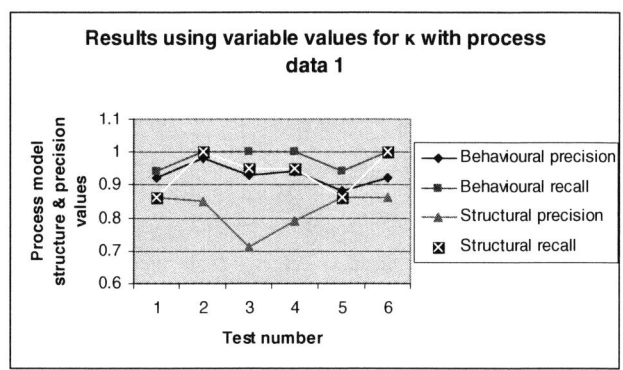

Fig. 5. Results using variable values for κ (process data 1)

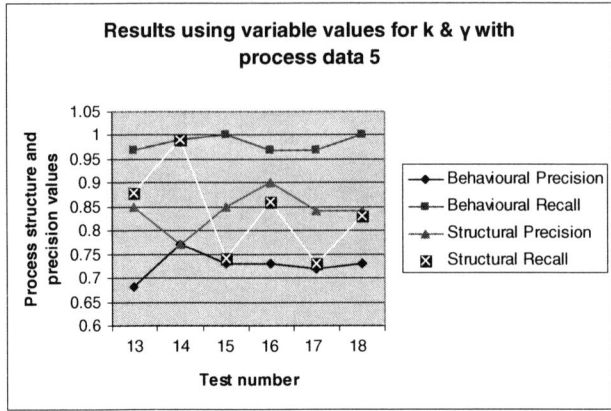

Fig. 6. Results using variable values for κ & γ with process data 5

5.2 Results of Mutation Rate Decay Experiments

The use of the decaying rate for the mutation operator was most noticeable in the results for process data 1. The use of the decay operator with mutation rate set initially between the values of .2 and .3 benefited the mining of the process model. The settings of 0.2, 0.21, 0.25, 0.27 and 0.28 allowed for the correct mining of the process model (shown in Fig. 7). This was an improvement over the results when using a constant rate of mutation (shown in Fig. 8). The use of the decaying rate was less noticeable for the remaining process data sets. The process model in process data 3 could be mined correctly regardless of the mutation rate. The effect of a decaying mutation rate on test 4 was marginal, with little or no difference to the results obtained without the decaying rate. Test 5 results, similarly did not seem to benefit from the decaying mutation rate. The mining of simple processes such as process data 1 benefited from the decaying rate of mutation.

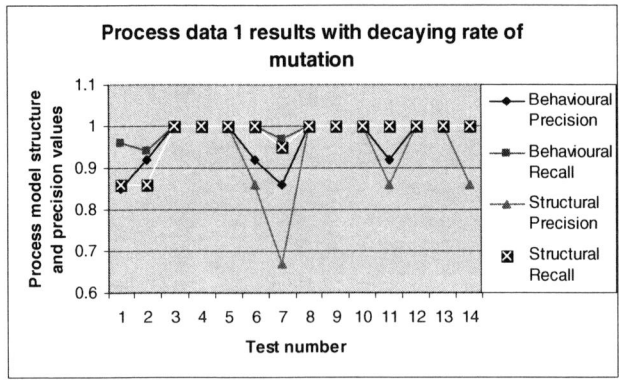

Fig. 7. Results for decaying rate of mutation with process data 1

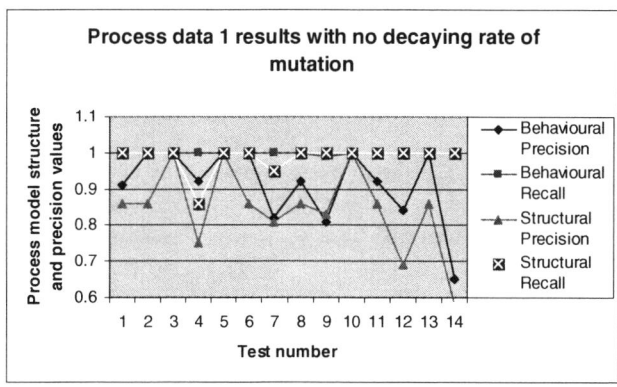

Fig. 8. Results for constant rate of mutation with process data 1

As mentioned earlier in section 4 of this paper the rate of mutation in the work of Alves de Mederios et al. [1] is set at a relatively high 0.2, due to the need to create non local causal relations. However it is likely that the continually high setting may actually hinder the achievement of a correct model in the later generations of the genetic algorithm. The effect of a decaying rate of mutation, or even a slightly higher or lower than recommended rate for mutation (with no decay) showed no real benefit in the more complex models such as process data 4 and 5.

6 Discussion

The mining of simple processes with a low level of loops (process data 1) bene-fited from a slightly higher value for the γ and a slightly lower value for the κ settings of the fitness function. In effect a higher weighting for γ decreases the likely-hood of process models with duplicates being selected for crossover, in this case leading to an improved mining result for process data 1. The lower setting of κ also helped to reintroduce a number of low frequency causal relations between certain activities (more low frequency relations were allowed in the process mod-els with a low κ setting), again aiding the correct mining of process data 1. Process data 4 also benefited from a slightly higher than recommended value for γ and a lower value for κ (with the other of the values set to 0.25). However the varying of the two parameters together did not bring noticeable benefits. Process data 2 seemed to benefit from a relatively high γ setting of 0.29 in combination with the recommended 0.25 setting for κ. Again the reduction in duplicates promoted by a higher γ setting promoted the correct rendering of the two loop constructs instead of their interpretation as four distinct tasks. While the experiments conducted in this paper have brought useful conclusions for the mining of simple processes containing low levels of loops, short sequences of tasks, and low levels of parallel-ism; further tests need to be conducted into the relationship between the operators in the genetic algorithm.

7 Conclusions

It is clear that an optimum general setting for the κ and γ parameters of the fitness function for use in all mining activities is not realistic (the κ parameter reduces the likelihood of process models with extra behaviour being selected and similarly the γ parameter restricts the selection of models containing duplicate tasks). The use of a higher value for the γ setting of the fitness function was shown to bring benefits for the mining of relatively simple processes. It is likely that a wider range of variables, such as crossover, population size and number of generations, need to be considered in the mining of more complex processes such as those provided by process data 4 and 5. However it may be possible to derive recommended settings for different classes of process. This conclusion can also be drawn for the mutation setting. The use of the decaying rate of mutation was shown to be beneficial in the mining of simple processes with a short sequence of tasks and few loops.

References

1. de Medeiros, A.K.A.: Genetic Process Mining. Ph. D Thesis, Eindhoven Technical University, Eindhoven, The Netherlands (2006)
2. de Medeiros, A.K.A., Weijters, A.J.M.M., van der Aalst, W.M.P.: Using Genetic Algorithms to Mine Process Models: Representation, Operators and Results. In: WP 124. Eindhoven Technical University, Eindhoven (2004)
3. Cox, E.: Fuzzy Modelling and Genetic Algorithms for Data Mining and Exploration. Morgan Kaufmann, London (2005)
4. Negnevitsky, M.: Artificial Intelligence. Addison Wesley, London (2005)
5. van Dongen, B.F., de Medeiros, A.K.A., Verbeek, H.M.W., Weijters, A.J.M.M., van der Aalst, W.M.P.: The ProM Framework: A new era in process mining tool support. In: Ciardo, G., Darondeau, P. (eds.) 26th International Conference on Applications and Theory of Petri Nets (ICATPN 2005). Springer, Heidelberg (2005)

## Part VII

Traffic and Transportation Systems

An Adaptive Differential Evolution Algorithm Applied to Highway Network Capacity Optimization

Andrew Koh

Institute for Transport Studies, University of Leeds, 39 University Road,
Leeds LS2 9JT, United Kingdom
{a.koh@its.leeds.ac.uk}

Abstract. Differential Evolution (DE) is a simple heuristic for global search. However, it is sensitive to certain control parameters and may not perform well if these parameters are not adjusted to suit the problems being optimized. Recent research has reported on methods to endogenously tune these control parameters during the search process. In this work, we develop and apply two DE variants as solution algorithms for continuous network design problems and illustrate with examples from the highway transportation literature.

Keywords: Continuous Network Design Problem, Differential Evolution, Bi-Level Programming.

1 Introduction

The continuous network design problem (CNDP) involves the determination of capacity enhancements, measured as continuous variables, of existing facilities of a network in such a way that the decision is regarded as optimal [1]. This remains a challenging research area within transportation [2].

In the literature, Genetic Algorithms (GA) [3] and Simulated Annealing (SA) [4] have been used to tackle the CNDP. Such stochastic search algorithms are capable of providing globally optimal solutions for many multi-modal optimization problems such as those encountered in the CNDP and its variants.

In this paper, we investigate two variants of Differential Evolution (DE) [5], a stochastic search heuristic, as solution algorithms for the CNDP. DE has already been used successfully in a multitude of applications to solve real world engineering problems. Like other stochastic search techniques, however, the performance of DE is susceptible to the choice of certain user defined control parameters. These control parameters are problem dependent; while one set of parameters may work well on some problems, they may not perform as well on others [6]. Thus, significant resources have to be devoted to adjusting these parameters. Hence many researchers have proposed adaptive DE algorithms where these control parameters themselves are evolved during the optimization process [6,7,8,9].

This paper is organized as follows. Following this introduction, Section 2 provides an overview of the CNDP. Section 3 then reviews the DE algorithm and a variant that allows endogenous parameter adaptation. Section 4 illustrates the performance of DE on two CNDPs from the highway transportation literature and finally Section 5 offers some conclusions and directions for further research.

E. Avineri et al. (Eds.): Applications of Soft Computing, ASC 52, pp. 211–220.
Springerlink.com   © Springer-Verlag Berlin Heidelberg 2009

2 Continuous Network Design Problem (CNDP)

The CNDP can be categorized as a Mathematical Program with Equilibrium Constraints ("bi-level programs"). These are mathematical programming problems with an equilibrium condition inextricably embedded in its formulation. Such programs are equivalent to Stackleberg or leader-follower games in which the leader chooses her decision variables so as to optimize her objective, taking into account the response of the followers who optimize their separate objectives [10].

In this game, the leader is the network planner/regulator and the followers are the network users. The users treat the planner's decision on capacity as exogenous when deciding their route choice. The usual assumption in the CNDP is that the route choice for a given level of capacity enhancement is based on Wardrop's user equilibrium principle [11] where user equilibrium is attained when no user can decrease his travel costs by unilaterally changing routes.

The difficulty with the bi-level program is that the leader cannot optimize her objective without considering the reactions of the followers. Even when both the leader's problem and the follower's problem separately consist of convex programming problems, the resulting bi-level problem itself may be non-convex [2]. Non convexity suggests the possibility of multiple local optima. Furthermore, additional capacity can counter-productively increase the total network travel time; this is the well-known Braess's Paradox [12]. Hence attention must be paid also to the network effects of providing additional link capacity which road users do not consider in their route choice decisions.

2.1 Model Formulation

Let:

$N : (X, A)$ represent the transportation network with X nodes and A links

R : the set of all routes in the network

H : the set of all Origin Destination (OD) pairs in the network

R_h : the set of routes between OD pair $h (h \in H)$

D_h : the demand between each OD pair $h (h \in H)$

f_r : the flow on route $r (r \in R)$

v : the vector of link flows, $v = [v_a]$ $(a \in A)$

$t_a(v_a)$: the travel time on the link a, as a function of link flow v_a on that link only.

δ_{ar} : 1 if the route $r (r \in R)$ uses link $a (a \in A)$, 0 otherwise

K : the set of links that have their individual capacities enhanced. ($K \subseteq A$)

β : the vector of capacity enhancements $\beta = [\beta_a], a (a \in K)$

$\beta_a^{max}, \beta_a^{min}$: the upper and lower bounds of capacity enhancements $a (a \in K)$

d_a : the monetary cost of capacity increments per unit of enhancement $a (a \in K)$

C_a^0 : existing capacity of link a $(\forall a \in A)$

θ : conversion factor from monetary investment costs to travel times

The CNDP seeks a K dimension vector of capacity enhancements that is optimal to the following bi-level program:

The Upper level problem (Program U) is given by

$$\underset{\beta}{\text{Min }} U(v,\beta) = \sum_{\forall a \in A} v_a(\beta) t_a(v_a(\beta), \beta_a) + \sum_{\forall a \in K} \theta d_a \beta_a \tag{1}$$

Subject to

$$\beta_a^{min} \le \beta_a \le \beta_a^{max} \quad \forall a \in K \tag{2}$$

Where v is obtained by solving the following lower level problem (Program L):

$$\underset{v}{\text{Min }} L = \sum_{\forall a} \int_0^{v_a} (t_a(z, \beta_a)) dz \tag{3}$$

Subject to

$$\sum_{r \in R_h} f_r = D_h, h \in H \tag{4}$$

$$v_a = \sum_{r \in R} f_r \delta_{ar}, \forall a \in A \tag{5}$$

$$f_r \ge 0, \forall r \in R \tag{6}$$

Program U defines the decision maker's objective as the sum of network travel times and investment costs of link capacity enhancements while Program L determines the user equilibrium flow, for a given β, based on Wardrop's first principle of route choice [11], formulated as an equivalent minimisation problem [13]. With β fixed, Program L can be solved via a traffic assignment algorithm.

The CNDP has been investigated by many researchers and various solution algorithms have so far been proposed. These have included Augmented Lagragian (AL) marginal function method [14] and Karush Khun Tucker (KKT) approaches [15]; both of which are derivative based methods. An approximation to Program U for an assumed β can be derived and direct search (DS) heuristics (i.e. search techniques that do not require derivatives such as golden section search) iteratively applied to approximately solve the CNDP [16]. Stochastic optimization techniques have also been used; GAs were applied in [17] and the use of SA has been reported in [1].

3 Differential Evolution Based Algorithms for CNDP

Differential Evolution (DE) is a multi-population based search heuristic [5] and has already been applied to a variety of real-world problems [18-20].

Our solution method using DE is as follows: At each iteration ("generation" in DE parlance), DE manipulates a population that comprises vectors of capacity enhancements (β) which are used to solve Program L, to obtain the link flows (v) to evaluate

$U(v, \beta)$ and determine the "fitness" of a given β. A fitter vector implies a lower value for $U(v, \beta)$ since we aim to minimize $U(v, \beta)$ (equation 2). This population is then transformed via DE operations (discussed herein) to create a new population with improved fitness and the entire process is repeated.

In this section, we outline two variants of DE. The first, which we refer to as "Basic DE" is the original DE version in [5] which operates with user specified control parameters. The second, which we refer to as "Adaptive DE", endogenously tunes these control parameters. Table 1 shows pseudo code of the two DE variants. The processes of initialization, evaluation, mutation, crossover and selection are common to both variants and discussed next.

Table 1. Comparison of Basic DE and Adaptive DE procedures in pseudo code

Procedure Basic DE	Procedure Adaptive DE
Generation = 1	Generation = 1
Initialization	Initialization
Evaluation	Evaluation
REPEAT	REPEAT
Mutation	*Create Control Parameters*
Crossover	Mutation
Evaluation	Crossover
Selection	Evaluation
UNTIL(Generation = MaxG)	Selection
	Update Control Parameters
	UNTIL (Generation = MaxG)

Initialization
An initial population of size= NP capacity enhancement vectors, known as the parent population in DE parlance, is randomly generated using equation 7 as follows:

$$\beta_{i,a,G} = rnd(\beta_a^{max} - \beta_a^{min}) + \beta_a^{min}, \forall i \in \{1, 2, ..., NP\}, \forall a \in K. \qquad (7)$$

rnd is a pseudo-random number$\in [0,1]$.

Evaluation
Each member of the population is a K dimension vector of capacity enhancements (β). The evaluation process involves solving Program L to determine the resulting link flows and enables evaluation of Program U for each member of this population; the member that results in the lowest objective function value for $U(v_a, \beta_a)$ is denoted the "best member" of the population ($\beta_{a,G}^{Best}$) at generation G.

Mutation
The mutation process combines different elements of the parent population heuristically to generate a mutant vector ($m_{i,a,G}$) in accordance with equation 8:

$$m_{i,a,G} = \beta_{i,a,G} + Q_i(\beta_{a,G}^{Best} - \beta_{i,a,G}) + Q_i(\beta_{a,G}^{r1} - \beta_{a,G}^{r2})$$ (8)

$$\forall i \in \{1,2,...,NP\}, \forall a \in K$$

$r1, r2 \in \{1,2,...,NP\}$ are random integer and mutually different indices and also different from the current running index i. Q_i is a mutation factor that scales the impact of the differential variation. The mutation strategy shown in equation 8 is one of several variants proposed in [5].

Crossover
On this mutant vector ($m_{i,a,G}$) crossover is probabilistically performed to produce a child vector ($y_{i,a,G}$) according to equation 9 as follows:

$$y_{i,a,G} = \begin{cases} m_{i,a,G} & \text{if } rnd \in [0,1] < CR_i \ \lor \ i = h \\ \beta_{i,a,G} & \text{otherwise} \end{cases} \quad \forall i \in \{1,2,...,NP\}, \forall a \in K$$ (9)

$h \in \{1,2,...,K\}$: a random integer parameter index chosen to ensure that the child vector $y_{i,a,G}$ will differ from its parent by at least one parameter. CR_i is the probability of crossover.

Crossover can produce child vectors that do not satisfy bound constraints in equation 2. Out of bound values can be reset to a point half way between its pre-mutation value and the bound violated using equation 10 as suggested in [21].

$$y_{i,a,G} = \begin{cases} \dfrac{\beta_{i,a,G} + \beta_a^{min}}{2} & \text{if } y_{i,a,G} < \beta_a^{min} \\[2mm] \dfrac{\beta_{i,a,G} + \beta_a^{max}}{2} & \text{if } y_{i,a,G} > \beta_a^{max} \\[2mm] y_{i,a,G} & \text{otherwise} \end{cases}$$ (10)

Selection
Each child vector is compared against the parent vector. This means that comparison is against the same i^{th} ($\forall i \in \{1,2,...,NP\}$) vector parent on the basis of whichever of the two gives a lower value for Program U . Using this selection procedure also prevents the occurrence of Braess's Paradox in using the DE based algorithm since only when the objective of Program U is reduced would it be regarded as fitter. The one that produces a lower value survives to become a parent in the next generation as shown in Equation 11.

$$\beta_{i,G+1} = \begin{cases} y_{i,G} & \text{if } U(v_a(y_{i,G}), y_{i,G}) < U(v_a(\beta_{i,G}), \beta_{i,G}) \\ \beta_{i,G} & \text{otherwise} \end{cases}$$ (11)

$$S_{i,G} \quad : 1 \text{ if } \beta_{i,G+1} = y_{i,G} \text{ ,0 otherwise} \tag{12}$$

For basic DE the procedures continue until some pre-specified number of generations ($MaxG$) are over. In Equation 12 $S_{i,G}$ is a dummy that takes on the value of 1 if the child vector survives or 0 otherwise. Equation 12 provides the link to Adaptive DE as will be shown next.

3.1 Adaptive DE

In basic DE it is conventionally further assumed [5] that:

$$Q_i = Q \ \forall i \in \{1, 2, ..., NP\} \quad and \quad CR_i = CR \ \forall i \in \{1, 2, ..., NP\} \tag{13}$$

In other words, in basic DE, it is assumed that the mutation and crossover factors are scalars. Thus basic DE requires 4 user-specified control parameters viz, Q , CR , NP and $MaxG$. There are some suggested values for these parameters. For example values such as $Q = 0.5, CR = 0.9$ have been suggested in [5,21]. However, in practice, many trial runs are required to find optimal parameters for each problem setting. Research developing adaptive versions tend to direct efforts at ways to endogenously compute Q and CR , leaving NP and $MaxG$ to remain user defined [6,7,8,9]. The latest contribution to a growing literature on parameter adaptation is in [9] which proposed the adaptive variant we discuss here. Other Adaptive DE versions can be found in [6,7,8].

The first point of departure for adaptive DE is to dispense with the assumption in Equation 13. This is done by separately associating each member of the population with its own crossover probability and mutation factor. Then, as an extension of equation 12, we may define the following:

$$S_{Q,G} = Q_i \cup S_{Q,G} \text{ if } S_{i,G} = 1 \text{ , } S_{Q,G} = S_{Q,G} \text{ otherwise.} \tag{14}$$

$$S_{CR,G} = CR_i \cup S_{CR,G} \text{ if } S_{i,G} = 1 \text{ , } S_{CR,G} = S_{CR,G} \text{ otherwise} \tag{15}$$

In other words, $S_{Q,G}$ and $S_{CR,G}$ denote the set of successful mutation factors and crossover probabilities used at generation G . Let the mean of these sets be denoted $\mu_{Q,G}$ and $\mu_{CR,G}$ respectively. Adaptive DE differs from Basic DE only in the creation and adaptation of the control parameters; steps shown in italics in the right pane of Table 1. All other processes are the same in both variants. The question then is how to obtain the vector of mutation and crossover factors for use at each generation in equations 8 and 9. We describe the methodologies to do so next.

Create Control Parameters
In addition to the initialization of a population of capacities, the user also specifies initial values of $\mu_{Q,G} = 0.7$ and $\mu_{CR,G} = 0.5$ ($G = 1$). Then production of the control parameters were suggested as follows [9]:

- generating $CR_i \ \forall i \in \{1, 2, ..., NP\}$ randomly from a normal distribution with mean $\mu_{CR,G}$ and standard deviation 0.1 truncated to be real numbers between 0 and 1;

- generating one third of Q_i $\forall i \in \{1,2,...,\frac{1}{3}NP\}$ randomly from a rectangular distribution as real numbers between 0 and 1.2;
- and generating two thirds of Q_i $\forall i \in \{\frac{1}{3}NP+1,\frac{1}{3}NP+2,...,NP\}$ randomly from a normal distribution with mean $\mu_{Q,G}$ and standard deviation 0.1 truncated to be real numbers between 0 and 1.2.

Update Control Parameters
Following mutation, crossover and selection, the following steps are carried out to update $\mu_{Q,G}$ and $\mu_{CR,G}$ for the next generation

1. Compute the Lehmer mean (Lh) of $S_{Q,G}$ using equation 16.

$$Lh = \frac{\sum\limits_{Q \in SQ,G} Q^2}{\sum\limits_{Q \in SQ,G} Q} \tag{16}$$

2. Update $\mu_{Q,G+1}$ using equation 17.

$$\mu_{Q,G+1} = (1-c)\mu_{Q,G} + cLh, \text{ where } c \in (0,1) \text{ is a user specified constant} \tag{17}$$

3. Compute Arithmetic mean ($\overline{S_{CR}}$) of $S_{CR,G}$
4. Update $\mu_{CR,G+1}$ using equation 18 as follows:

$$\mu_{CR,G+1} = (1-c)\mu_{CR,G} + c\overline{S_{CR}} \text{ where } c \text{ is as used in 17} \tag{18}$$

Once the means are updated, the generation method can be applied to create CR_i and Q_i parameters for the mutation and crossover processes. In summary, the main difference between adaptive DE and basic DE is the use of vector based control parameters; the parametric distributions used to generate these are iteratively adapted in the algorithm depending on the occurrence of successful selection.

4   Numerical Examples

We report on the performance of basic DE and adaptive DE based algorithms for solving the CNDP with other reported solutions on two test problems. In addition, we used the Origin Based Assignment algorithm [22] to solve the Program L during the evaluation phase of DE. Our reported results for the DE methods are the average upper level problem objective values i.e. $U(v,\beta)$ and its standard deviation (SD) over 30 runs. The results for other algorithms are taken directly from the cited sources.

4.1 Example 1

The CNDP for the hypothetical network of 16 links and 2 OD pairs is used as the first example. This network, its parameters and trip details are taken from [16]. All 16 links were subject to capacity enhancements with $\beta_a^{min} = 0$, $\beta_a^{max} = 20$ (for all links) and $\theta = 1$. For basic DE, we assumed Q and CR to be 0.8 and 0.95 respectively. For adaptive DE, we assumed c to be 0.01. In both cases $NP = 20$ and $MaxG = 150$. The results are shown in Table 2. Note that the gradient based methods [14, 15] can be mathematically shown to converge at least to a local optimum while both of our DE based methods are heuristics. As the CNDP is a non-convex problem [2], the ability of these former methods to locate the global optimum is dependent on the starting point assumed.

Table 2. Comparison of DE variants with other approaches for Example 1

Method	KKT	AL	SA	GA	Basic DE	Adaptive DE
Source	[15]	[14]	[1]	[17]		
Objv	534.02	532.71	528.49	519.03	522.71	523.17
NEval	29	4,000	24,300	10,000	3,000	3,000
SD	------------Not Available-----------			0.403	1.34	0.97

Note: Objv: value of $U(v, \beta)$ at end of run; NEval: number of Program L (traffic assignments) solved.

4.2 Example 2

The second example is the CNDP for the Sioux Falls network with 24 nodes, 76 links and 552 OD pairs. The network and travel demand details are found in [16]. 10 links out of the 76 are subject to capacity enhancements; $\beta_a^{min} = 0$, $\beta_a^{max} = 25$ (for all the 10 links) and θ is 0.001. For basic DE, we assumed Q and CR to be 0.8 and 0.9 respectively. For adaptive DE, c was 0.01. In both cases $NP = 20$ and $MaxG = 80$. The literature does not indicate that GA has been used for this problem. The results are shown in Table 3.

Table 3. Comparison of DE variants with other approaches for Example 2

Method	DS	KKT	AL	SA	Basic DE	Adaptive DE
Source	[16]	[15]	[14]	[1]		
Objv	83.08	82.57	81.75	80.87	80.74	80.74
NEval	12	10	2,000	3,900	1,600	1,600
SD	---------------Not Available--------------				0.002	0.006

Note: Objv: Value of $U(v, \beta)$ at end; NEval: number of Program L (traffic assignments) solved.

Note that while this network is clearly larger and arguably more realistic, the problem dimension (number of variables optimized) is smaller than in Example 1, since 10 links are subject to improvement compared to 16 in Example 1. This could explain why the number of Program L problems solved are less than in Example 1.

5 Conclusions

We developed and applied a basic DE and adaptive variant as solution heuristics for the CNDP. In our numerical tests, we applied these methods to a hypothetical 16 link, 2 OD pair network and the Sioux Falls network with 76 links and 552 OD pairs.

As a global optimization heuristic, it has been concluded [5] that DE is competitive with SA and GA. Our results support this view as the DE-based methods required a lower number of function evaluations to obtain/better the optimum found by GA and SA. The GA based [17] results in example 1 are better but at the expense of extensive function evaluations. In future work, DE could be hybridized with local search algorithms to home in on the global optimum as in [6]. Our results for the Sioux Falls network in example 2 are arguably better than any in the literature so far.

Our results are also competitive against various derivative-based methods. Compared to the KKT methods in [15], DE required many more function evaluations but DE obtained a lower function value, suggesting DE managed to escape a local optimum. On the other hand, the gradient based AL method [14] required more function evaluations than the DE based methods but did not reach the optimum value.

The low variance augments the view that the DE methods are also reasonably robust. Hence this implies that the DE based method should be able to consistently locate the region of the optimum in multiple trials. Furthermore, a population size NP much less than suggested in [5] was used to obtain the results shown in this paper.

It is difficult to compare the performance of Basic and Adaptive DE since both could provide solutions that are quite similar. Furthermore, the standard deviations of both are not too different. Nevertheless, we point out that we carried out extensive initial testing to decide the mutation and crossover parameters ultimately used for the Basic DE algorithm which attests to the primary advantage of the adaptive version.

The downside of the adaptive variant is that further sensitivity analysis of the c parameter used for updating the crossover and mutation factors, needs to be carried out to examine its robustness with different test problems although [9] suggests that it is not critical. We conjecture that this parameter might be problem dependent. Further research is required to investigate this adaptive DE variant before firm conclusions can be reached.

Another potential avenue for further research could be to apply such stochastic search heuristics to multi-objective optimization CNDPs where tradeoffs between objectives need to be examined.

Acknowledgments. The work reported forms research funded by UK EPSRC. The author thanks Prof Hillel Bar-Gera (Ben Gurion University of the Negev, Israel) for provision of his Origin Based Assignment software. Comments on earlier drafts were received from Prof Mike Maher. Errors remain the author's sole responsibility.

References

1. Friesz, T., Cho, H., Mehta, N., Tobin, R., Anandalingam, G.: A Simulated Annealing Approach to the Network Design Problem with Variational Inequality Constraints Transport. Sci. 26, 18–26 (1992)
2. Bell, M., Iida, Y.: Transportation Network Analysis. John Wiley, Chichester (1997)
3. Goldberg, D.E.: Genetic Algorithms in Search, Optimization, and Machine Learning. Addison-Wesley, Reading (1989)
4. Aarts, E., Korst, J.: Simulated Annealing and Boltzman Machines. John Wiley, Chichester (1988)

5. Storn, R., Price, K.: Differential Evolution – A Simple and Efficient Heuristic for Global Optimization over Continuous Spaces. J. Global Optim. 11, 341–359 (1997)
6. Qin, A., Suganthan, P.N.: Self-adaptive differential evolution algorithm for numerical optimization. In: Proceedings of 2005 IEEE Congress on Evolutionary Computation, pp. 1785–1791 (2005)
7. Brest, J., Bošković, B., Grenier, S., Žumer, V., Maučec, M.V.: Performance comparison of self-adaptive and adaptive differential evolution algorithms. Soft Comput 11, 617–629 (2007)
8. Liu, J., Lampinen, J.: A Fuzzy Adaptive Differential Evolution Algorithm. Soft Comput 9, 448–462 (2005)
9. Zhang, J., Sanderson, A.: JADE: Self-Adaptive Differential Evolution with Fast and Reliable Convergence Performance. In: Proceedings of 2007 IEEE Congress on Evolutionary Computation, pp. 2251–2258 (2007)
10. Fisk, C.: Game theory and transportation systems modeling Transport. Res. B-Meth. 18, 301–313 (1984)
11. Wardrop, J.G.: Some theoretical aspects of road traffic research. Proc. I. Civil Eng.-Tpt. 1, 325–378 (1952)
12. Braess, D.: Uber ein paradoxon aus der verkehrsplanung. Unternehmenforschung 12, 258–268 (1968)
13. Beckmann, M., McGuire, C.B., Winsten, C.B.: Studies in the Economics of Transportation. Yale University Press, New Haven (1956)
14. Meng, Q., Yang, H., Bell, M.: An equivalent continuously differentiable model and a locally convergent algorithm for the continuous network design problem. Transport. Res. B-Meth. 35, 83–105 (2000)
15. Chiou, S.: Bilevel programming formulation for the continuous network design problem. Transport. Res. B-Meth. 39, 361–383 (2005)
16. Suwansirikul, C., Friesz, T., Tobin, R.L.: Equilibrium decomposed optimization, a heuristic for continuous network design. Transport. Sci. 21, 254–263 (1997)
17. Cree, N.D., Maher, M.J., Paechter, B.: The continuous equilibrium optimal network design problem: a genetic approach. In: Bell, M. (ed.) Transportation Networks: Recent Methodological Advances, Pergamon, London, pp. 163–174 (1996)
18. Ilonen, J., Kamarainen, J., Lampinen, J.: Differential Evolution Training Algorithm for Feed-Forward Neural Networks. Neural Process. Lett. 7, 93–105 (2003)
19. Karaboga, N.: Digitial IIR Filter Design Using Differential Evolution Algorithm. Eurasip J. Appl. Si. Pr. 8, 1269–1276 (2005)
20. Qing, A.: Dynamic differential evolution strategy and applications in electromagnetic inverse scattering problems. IEEE Trans. Geosci. Remote Sens. 44, 116–125 (2006)
21. Price, K.: An Introduction to Differential Evolution. In: Corne, D., Dorigo, M., Glover, F. (eds.) New Techniques in Optimization, pp. 79–108. McGraw Hill, London (1999)
22. Bar-Gera, H.: Origin based algorithm for the traffic assignment problem. Transport. Sci. 36, 398–417 (2002)

Incorporating Fuzzy Reference Points into Applications of Travel Choice Modeling

Erel Avineri

Centre for Transport & Society, University of the West of England, Bristol, BS16 1QY, UK

Summary. The uncertainties involved in travel alternatives have an effect on travelers' choices, however travel choice models and commercial applications have limitations in capturing the uncertainty in the mind of the travelers. Travelers' preferences, as revealed in field studies and laboratory studies, are generally supported by the robust findings of Prospect Theory, a descriptive model of individual choice under risk and uncertainty. Prospect Theory models responses to risky situations framed as 'gains' and 'losses' defined over a reference point. However, different from choices made in economic/financial contexts, the concepts of 'winning' or 'loosing' in a travel choice context may be considered to be fuzzy rather than crisp. Extending the principles of prospect theory, this paper introduces a model based on a fuzzy representation of the travel time's reference point in the mind of the traveler.

1 Introduction

Travel choice models are designed to emulate the behavior of travelers over time and space and to predict changes in system performance, when influencing conditions are changed. Such models include the mathematical and logical abstractions of real-world systems implemented in computer software. Uncertainties in the transport system (such as the reliability of transport modes) affect the travel choices made by individuals. Traditionally, uncertainties of transport systems have been addressed in research works and commercial tools (such as equilibrium models, discrete choice models or micro-simulations) by probability measures. It has been argued that common models of travelers' behavior and commercial applications do not offer much insight on the way travelers make choices. In applications that do address the uncertainty or variability in the transport system, travel choice models used mainly measure the uncertainty of the system, and do not attempt to capture the uncertainty in the mind of the traveler, and its effect on travel choices. In order to make practical use of travelchoice models in stochastic networks a link is required between objectively measurable uncertainty of the transport system and travelers' perception of that uncertainty. The aim of this study is to combine elements of travelers' responses to uncertainty in order to represent travel choice in complex travel behavior contexts. In this work it is suggested to combine three important aspects of

E. Avineri et al. (Eds.): Applications of Soft Computing, ASC 52, pp. 221–229.
springerlink.com

uncertainty: (i) stochasticity of the possible outcomes of a travel choice (i.e. each choice's distribution of travel time); (ii) framing stochastic outcomes as positive ('gains') or negative ('losses') from the agent's perspective, in relation to a reference point; and (iii) representing travel time's reference point, as perceived by the individual making a travel choice, as a fuzzy notion. Combining the above elements in modeling travelers' responses to uncertainty is illustrated by a travel choice model developed in this work. A numeric example to illustrate the application is given.

2   Prospect Theory and Its Applications in Modeling Travel Choice Behavior

For decades, Expected Utility Theory (EUT) [1] has been the dominant approach in modeling individuals' decision making under risk and uncertainty. For example, it is common to measure the performance of a transport system by the trade off between travel time (or other attributes) and its variability. Both travel time and its variability are commonly considered as negative utilities to the traveler, who is mainly interested in making short and reliable journeys. Many EUT models of travel behavior are based on the assumption that travelers are utility maximizers and risk averse that seek to trade off travel time and travel time variance explicitly.

Experiments in behavioral studies often find systematic deviations from the predictions of classical theory of EUT [2, 3]. Unlike traditional economic theories, which deduce implications from normative preference axioms, Prospect Theory (PT) [2, 3] takes a descriptive approach. PT assumes that probabilistic outcomes are mapped as gains or losses relative to some reference point. PT utilizes a value function $\nu(\cdot)$ and a probability weighting function $\pi(\cdot)$. For example, consider a lottery with three outcomes: x with a probability p, y with a probability q, and the status-quo with a probability $1 - p - q$. The prospect-theoretic value of the lottery is given by:

$$\pi(p)\nu(x) + \pi(q)\nu(y) \tag{1}$$

In PT, the carriers of utility, described by the value function, are gains and losses measured against some implicit (crisp) reference point. The value function v(.) is assumed to be concave in gains and convex in losses, a pattern which is consistent with the experimental evidence on domain-sensitive risk preferences [2, 3, 4, 5]. An empirical observation on individuals' preferences in risky environment is that people treat gains and losses differently. This observed risk-taking behavior, called *loss aversion*, refers to the fact that people tend to be more sensitive to decreases in their wealth than to increases. To capture loss aversion, the value function is assumed to have a stepper slope for losses than for gains. These features of the value function are presented in Fig. 1.

Experimental results [2, 3, 4, 5] also imply a reversed-S-shaped probability weighting function. Diminishing marginal sensitivity occurs in this function with

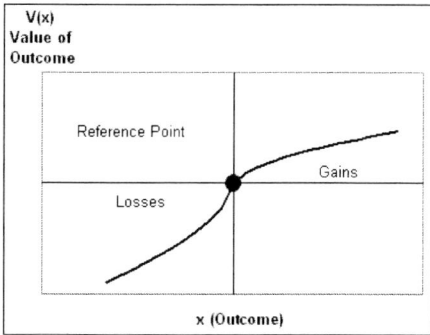

Fig. 1. PT Value Function

respect to the benchmark case of certainty. As probabilities move further away from the 0 and 1 end-points, the probability weighting function flattens out. Experimental results reveal that this curve tends to lie disproportionately below the 45 degrees line, as shown in Fig. 2 [2, 3, 4, 5].

A version of PT that employs cumulative rather than separable decision weights, called Cumulative Prospect Theory (CPT) [3] applies the cumulative functional separately to gains and to losses. An uncertain prospect f is a function from a finite set of states of nature S into a set of outcomes X, that assigns each state an outcome. To define the cumulative functional, the outcomes of each prospect are arranged in increasing order of their values. CPT asserts that there exist a strictly increasing value function $\nu : X \to \mathcal{R}$, satisfying $\nu(x_0) = \nu(0) = 0$, and decision weights functions w^+ and w^- such that,

$$CWV(f) = \nu(f^+) + \nu(f^-) \tag{2}$$

$$\nu(f^+) = \sum_{i=0}^{n} \pi_i^+ \nu(x_i) \tag{3}$$

$$\nu(f^-) = \sum_{i=-m}^{0} \pi_i^- \nu(x_i) \tag{4}$$

where $\nu(f^+)$ is the prospect gains value, $\nu(f^-)$ is the prospect losses value, $\pi^+(f^+) = (\pi_0^+, \cdots, \pi_n^+)$ are the decision weights of the gains, and $\pi^-(f^-) = (\pi_{-m}^-, \cdots, \pi_0^-)$ are the decision weights of the losses.

The following functional form for the value function fits the CPT assumptions:

$$\nu(x) = \begin{cases} x^\alpha & \text{if } x \geq 0 \\ -\lambda(-x)^\beta & \text{if } x < 0 \end{cases} \tag{5}$$

Tversky & Kahneman [3] proposed the following formulation of the weighting for gains and losses, respectively:

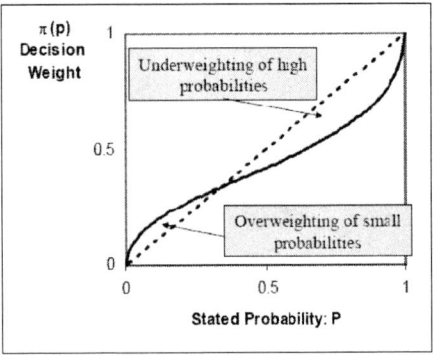

Fig. 2. PT Weighting Function

$$w^+(p) = p^y/[p^y + (1-p)^y]^{1/y} \tag{6}$$
$$w^-(p) = p^\delta/[p^\delta + (1-p)^\delta]^{1/\delta} \tag{7}$$

CPT has already been applied fruitfully to some settings in economics. Travel behavior researchers have recently studied the application of PT and CPT to route choice modeling, departure time modeling and passengers' choice of public transport service. Travelers' preferences, as revealed in field studies and laboratory studies reported in the above studies, are generally supported by the robust findings of CPT, assuming a crisp reference point. This paper explores a possible extension of the use of PT in travel choice modeling. We introduce the concept of a fuzzy rather than crisp representation of travel choice outcomes as 'gains' and 'losses', and a measure of fuzziness is used to represent traveler's vague perception of a reference point. We illustrate how this concept can be incorporated into the modeling of travel choice.

3   Reference Point: Crisp and Fuzzy Approaches

PT was originally developed to describe and predict risky choice in a financial/monetary context. Travelers, aware of the reliabilities of transport systems, make their route, mode or travel time choices in situations which are generally more complex than the decision making scenarios studied by experimental economics and economic psychologists, commonly framed as monetary-based gambling problems. A main difference between making a choice in a monetary/financial context and a travel-behavior context is the lack of consensus about the value of the reference point in the later one. $0 may be the common reference point monetary gains and losses. However, dealing with traveler's perceptions of travel time attributes, the framing of the experienced travel time as a gain or a loss is based on an individual reference point, which can be a mental or a verbal representation of the traveler's past experience and expectations. Such a reference point may differ from one traveler to another, may also differ from decision to decision, as it is updated over time as new information is gathered by travelers.

Within the context of travel behavior modeling, a reference point may represent a threshold value that distinguishes between 'gains' and 'losses' of the journey, as perceived by travelers. While we do not have an adequate model to predict (or estimate) the value of the reference point, it was found in many studies that humans' adoption of a reference point is influenced by implicit information, explicit information, or even irrelevant information. A-priori perception of travel time, influencing travelers' value of the reference point, may have an indirect influence on travelers' choices. In the absence of a methodology to evaluate the value of the reference point, one may assume that it is related to the actual travel time experienced (roughly about 20-30 minutes for many commuters).

The boundaries of representing the reference point should not necessarily be sharp. A fuzzy representation of the traveler's cognitive "soft" observation/experience of the reference point may yield more informative and reliable interpretation than does the traditional crisp representation of reference point. In the context of travel choice we suggest an alternative model of referencing. A traveler does not necessarily have a crisp and sharp definition of a reference point in mind, therefore a sound assumption may be that the perception of reference point in the mind of a traveler is vague or fuzzy rather than crisp.

Eq. 8 defines a (triangular) fuzzy reference point, $\tilde{RP} = (rp_1, rp_2, rp_3)$ where rp_1 is lower boundary of the triangular fuzzy reference point, rp_2 is the number corresponding to the highest level of presumption, and rp_3 is upper boundary of the fuzzy reference point.

$$\mu_{RP}(x) = \begin{cases} 0, & x \leq rp_1 \\ \left(\frac{x-rp_1}{rp_2-rp_1}\right), & rp_1 < x < rp_2 \\ \frac{rp_3-x}{rp_3-rp_2}, & rp_2 \leq x \leq rp_3 \\ 0, & rp_3 \leq x \end{cases} \tag{8}$$

In the context of travel choice, 'gains' and 'losses' are defined as follows: If $x \leq rp_1$ than x is in the gains side (where the journey is clearly taking less time than expected); if $x \geq rp_3$ than x is in the 'losses' side. For any x value between rp_1 and rp_3, x has positive membership values to both 'gains' and 'losses', as defined in Eqns. 9 and 10.

$$\mu_{gain}(x) = \begin{cases} 1 & x \leq rp_1 \\ 1 - \frac{x-rp_1}{rp_3-rp_1} & rp_1 < x < rp_3 \\ 0 & rp_3 \leq x \end{cases} \tag{9}$$

$$\mu_{loss}(x) = \begin{cases} 0 & x \leq rp_1 \\ \frac{x-rp_1}{rp_3-rp_1} & rp_1 < x < rp_3 \\ 1 & rp_3 \leq x \end{cases} \tag{10}$$

It can be easily shown that the crisp reference point is a special case of the fuzzy reference point, where $rp = rp_1 = rp_2 = rp_3$, and instead of fuzzy sets representation of 'gains' and 'losses', a crisp value, rp, provide a crisp distinction between the two regions.

The extension principle [6, 7] can be used in the generation of the mathematical crisp concept of CWV to a fuzzy set. The fuzzy set of CWV, $C\tilde{W}V$ is defined as follows:

$$C\tilde{W}V = g(X, \tilde{rp}) = \{(y, \mu_{C\tilde{W}V}(y)) \mid y = CWV(X, rp)\} \tag{11}$$

where

$$\mu_{C\tilde{W}V}(y) = \begin{cases} \sup_{rp \in g^{-1}(y)} \mu_{C\tilde{W}V}(X, rp) & \text{if } g^{-1}(y) \neq 0 \\ 0 & \text{otherwise} \end{cases} \tag{12}$$

A fuzzy graph [8] may provide an equivalent representation for $C\tilde{W}V$.

4   Numeric Example

Fig. 3 represents a situation of route choice under risk. The cumulative weighted values of route A and route B prospects (CWV_A and CWV_B, accordingly) can be estimated based on functional shapes and parameters estimated in [3]. For example, for a reference point of 25 minutes, the CPT values are $CWV_A = -3.8$ and $CWV_B = -5.1$, thus route A is more attractive than route B. The sensitivity of CPT predictions to reference point values directly results from the different slope ratios for gains and losses, on each side of the reference point.

Traveler perception of travel time's reference point may be fuzzy rather than crisp, thus a range of values $rp_1 \leq x \leq rp_2$ represents outcomes that have positive membership values to both 'gains' and 'losses'. Applying the extension of the mathematical crisp concept of CWV from crisp variable to a fuzzy set (Eq. 11-12), the fuzzy CPT predictions of route choice are illustrated by four situations: (i) Traveler's reference point of travel time is $RP = 25$ minutes (crisp); (ii) A fuzzy reference point of travel time $\tilde{RP} = (20, 25, 30)$; (iii) A fuzzy reference

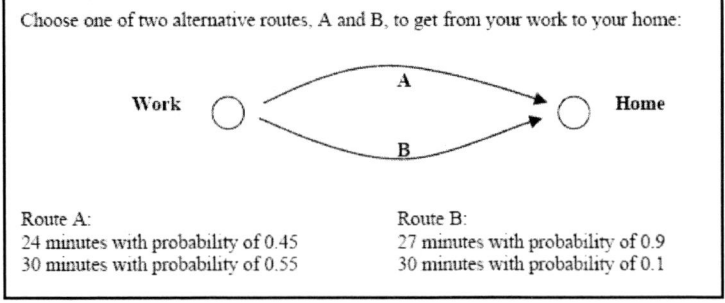

Fig. 3. Route Choice Problem

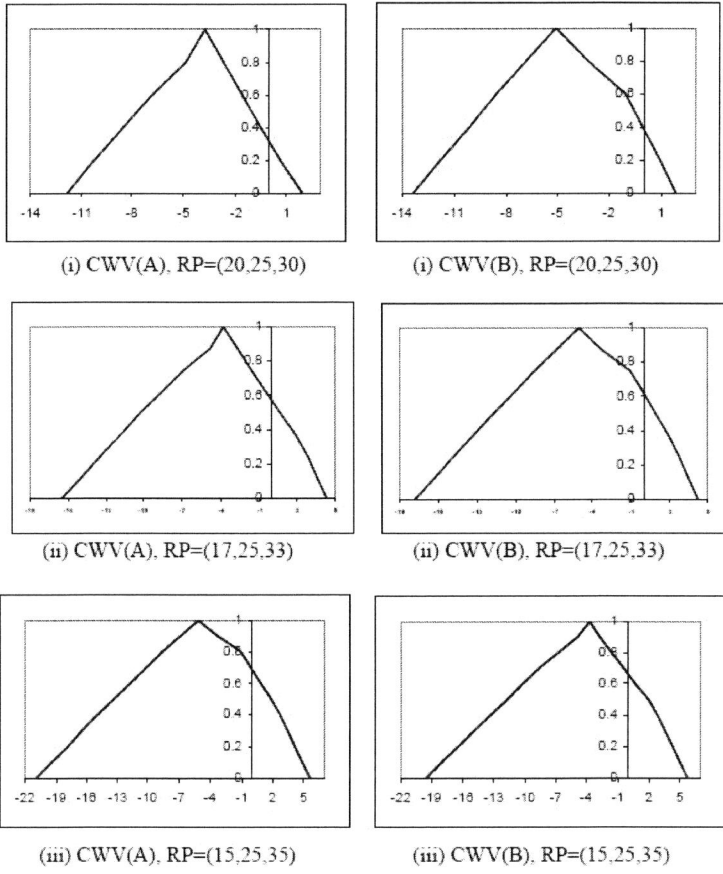

(i) CWV(A), RP=(20,25,30)   (i) CWV(B), RP=(20,25,30)

(ii) CWV(A), RP=(17,25,33)   (ii) CWV(B), RP=(17,25,33)

(iii) CWV(A), RP=(15,25,35)   (iii) CWV(B), RP=(15,25,35)

Fig. 4. Cumulative weighted values as fuzzy sets: 3 examples

point of travel time $\tilde{RP} = (17, 25, 33)$; (iv) A fuzzy reference point of travel time $\tilde{RP} = (15, 25, 35)$. Based the above definitions of membership functions (ii,iii,iv) and the extension of CPT definitions to address the fuzziness of the reference point (as described in section 3) the membership functions of the cumulative weighted values for each route prospect have been generated and are presented in Fig. 4.

Table 1 presents the defuzzified cumulative weighted values of route prospects (based on the Center of Area method), and the prediction of the route choice for each of the four situations. The distribution of cumulative weighted values around a crisp reference point is not symmetric. Thus, different shapes of the membership function $\tilde{RP}$ lead to different weightings of cumulative weighted values. For this reason, the cumulative weighted values of route prospects are highly sensitive to the fuzziness around the reference point, as can be seen from table 1.

Table 1. Cumulative weighted values and CPT predictions

	Reference Point	CWV_A	CWV_B	CPT Prediction
(*)	RP=25	-3.8	-5.1	$CWV_A > CWV_B$
(i)	RP=(20,25,30)	-5.8	-6.8	$CWV_A > CWV_B$
(ii)	RP=(17,25,33)	-6.8	-6.8	$CWV_A \approx CWV_B$
(iii)	RP=(15,25,35)	-6.9	-7.7	$CWV_A > CWV_B$

(*)-crisp CWV; (i),(ii),(iii)-defuzzified CWV.

As illustrated in Table 1, the fuzzier the reference point is, the less attractive the travel choice becomes (cumulative weighted values having higher negative values). This can be explained by the general feature of asymmetry of the value function around a reference point; setting a fuzzy rather than crisp value to a reference point, 'winning' crisp outcomes are replaced by both fuzzy gains and losses, making outcomes formally framed as gains to be less attractive.

5   Conclusion and Future Research

In this work, a combined fuzzy/prospect theory approach to model travelers' responses to uncertainty was developed. The aim of this work is to combine different types of uncertainty that play an important role in cognitive choice behavior: travelers' responses to probabilistic outcomes (described by prospect theory) and a fuzzy perception of gains and losses, related to a travel time's reference point. Travel choice gains and losses were defined over a fuzzy region of travel times instead of a single crisp reference.

In the context of travel choice, where a traveler does not necessarily have a crisp and sharp definition of a reference point in mind, a sound assumption may be that traveler's perception of reference point is vague or fuzzy. However, more empirical research on travelers' reference points and how they are formed needed to be done to validate the fuzzy reference point hypothesis and to gather understanding of its relevance in a travel choice context.

References

1. Von Neuman, J., Morgenstern, O.: Theory of Games and Economic Behavior. Princeton University Press, USA (1944)
2. Kahneman, D., Tversky, A.: Prospect Theory: An Analysis of Decisions under Risk. Econometrica 47, 263–291 (1979)
3. Tversky, A., Kahneman, D.: Advances in Prospect Theory: Cumulative Representation of Uncertainty. Journal of Risk and Uncertainty 9, 195–230 (1992)
4. Preston, M., Baratta, P.: An Experimental Study of the Auction-Value of an Uncertain Outcome. American Journal of Psychology 61, 183–193 (1948)
5. Gonzalez, R., Wu, G.: On the Form of the Probability Weighting Function. Cognitive Psychology 38, 129–166 (1999)

6. Zadeh, L.A.: The Concept of Linguistic Variable and its Application to Approximate Reasoning. Memorandum ERL-M 411, Berkeley (October 1973)
7. Dubois, D., Prade, H.: Fuzzy Sets and Systems: Theory and Applications. Academic Press, New York (1980)
8. Wolkenhauer, O.: Data Engineering: Fuzzy Mathematics in Systems Theory & Data Analysis Summary. John Wiley and Sons, Chichester (2001)

The Role of the Uncertainty in ATIS Applications

Gennaro Nicola Bifulco[1], Fulvio Simonelli[1], and Roberta Di Pace[2]

[1] University of Napoli Federico II, Department of Transportation Engineering Via
    Claudio 21, 80125 Napoli, Italy
    {gennaro.bifulco,fulsimon}@unina.it
[2] Delft University of Technology, Section Transportation and Planning
    P.O. Box 5048, 2600 GA Delft, The Netherlands
    R.diPace@student.tudelft.nl

Summary. In recent years the interest toward ATIS (Advanced Traveller Information Systems) is constantly increasing, probably because of the intent of solving over-
congestion with minimal expenditure. Great efforts have been devoted to both technological and modeling aspects; however, a unified theory does not yet exist, as well as a
general agreement on the objectives that can be achieved. It is not widely recognised
what can be expected from ATIS, how the information should be designed and how
to assess/forecast the network level effects. Here the authors will present a theoretical
framework and some practical analyses, aimed to deal with the additional uncertainties
sometimes introduced in the route choice problem by the presence of ATIS.

1 Introduction

ATIS represent an *additional* source of information for travellers. The informa-
tion/estimates of the traffic conditions that travellers already own because of
their familiarity with the traffic network can be improved by ATIS. In this context, static ATIS are of little interest for the great part of the travellers, except
for totally unfamiliar ones. Moreover, it has been shown in [1] that static ATIS
could decrease the ability of travellers to pursue their optimum route choice behaviours. It is more useful to provide dynamic information varying with traffic
conditions, this is aimed to describe traffic conditions (typically travel-times)
of a set of routes (*descriptive information*) or to suggest the best route-choice
(*prescriptive information*) on the base of ATIS-estimated travel times. The information is *instantaneous* when referred to instantaneous travel times or *predictive*
if referred to actual travel times. Instantaneous travel times at a given time are
the sum of all link travel-times that compose the routes. Actual travel times are
the ones that the travellers will have experimented at the end of their trip. Also
in the simpler case that ATIS supplies information on the base of instantaneous
travel times, some inaccuracies could result. These can depend, for instance, on
the fact that travel times are inferred from the ones monitored only on a part of
the links and, possibly, are not directly measured but inferred from other variables. When ATIS try to supply predictive information, the problem is even more

E. Avineri et al. (Eds.): Applications of Soft Computing, ASC 52, pp. 230–239.
springerlink.com   © Springer-Verlag Berlin Heidelberg 2009

complex, due to the presence of congestion. In fact, as the number of drivers that receive information (*market penetration rate*) and react to it (*compliance rate*, here referred both to the descriptive and prescriptive case, in accordance to [2]) increase, it becomes important in generating information to take account of the effects on traffic conditions ([3]). This is not only aimed to avoid effects such as overreaction and concentration, but also to ensure that the messages based on predicted traffic conditions remain consistent with those conditions. This is also known ([4]) as the anticipatory route guidance problem. It is hard to be exactly resolved for real networks in real (or fast) time and represents a further source of uncertainty. In summary, the information supplied by ATIS is intrinsically inaccurate and this is even more probable for predictive information. It is also worth noting that the static information is inaccurate by definition, at least in case of congested networks. This gives rise to uncertainty also with respect to the choice to be compliant to ATIS, where compliance is here intended as the propensity to take into account the ATIS and to react to it. This uncertainty interacts with the one of route travel times, which is due different levels of familiarity to the network and to the variability of traffic patterns and travel times. It is evident that if the route travel times have a very small dispersion over time and if all travelers are familiar to the network, the uncertainty of the ATIS information does not play a significant role, provided that the travelers will finally choose on the base of their own knowledge of the network. If, on the other hand, the uncertainty of the (non-ATIS) owned information is greater, the travelers will probably use and react to ATIS and the degree of compliance will be greater the smaller will be uncertainty. In this paper the uncertainty will be dealt with in the context of the probabilistic theory; other authors ([5]) have preferred to approach ATIS problems through the possibility theory. It is not the sake of this paper to discuss on the advantages or disadvantages of each approach (see for instance [6]) and the more consolidated probabilistic theory is here assumed as reference; moreover, we argue that the general results here presented should be also reached within a possibility approach.

2   Decision Making under Uncertainty in ATIS Contexts

The uncertainty has to be handled in the simulation of route choice behaviors also taking into account the extra-uncertainty due to ATIS. To this aim the authors suggest a conditional probability approach, framed within the random utility theory. The joint probability that a traveler is compliant (not compliant) and chooses a given route con be expressed as:

$$P(C, i) = P(C) \cdot P(i/C) = \eta \cdot \pi_i^Y$$
$$P(NC, i) = P(NC) \cdot P(i/NC) = (1 - \eta) \cdot \pi_i^N$$

(1)

Where: π_i^Y is the probability of choosing route i provided the traveler is compliant; π_i^N is the probability of choosing the same route provided the traveler is not compliant; η is the probability that traveler who has access to information (*equipped*) is compliant. The systematic utility of the generic alternative *(c, j)*, where *c* takes the values *c=C* and *c=NC* and *i* is the chosen route, depends on proper attributes (explicative variables), authors' opinion is that among these attributes a significant role should be played by tha ATIS accuracy. The (in)accuracy of the ATIS information can be defined in several ways (see for instance [4]); here we propose to use descriptive inaccuracy:

$$\text{Descriptive (in)accuracy } Dinc = \frac{\|G - C\|}{\|C\|} \tag{2}$$

The previous can be straight used in case of descriptive ATIS (to which we will refer in the following) and can be substituded by the descriptive inaccuracy in the case of prescriptive ATIS:

$$\text{Prescriptive (in)accuracy } Pinc = Abs\left(\frac{\Phi^T \cdot C - \Pi^T \cdot C}{\Pi^T \cdot C}\right) \tag{3}$$

In previous equations: *G* is the vector of route travel times supplied by the ATIS; *C* is the vector of actual travel times experienced by the travelers after their trips have been made; *Φ* is the vector of route choices percentages as suggested by the ATIS (in case of prescriptive ATIS) or (in case of descriptive ATIS) the vector of route choices percentages as it were resulted by the behavior of the travelers in correspondence to the ATIS supplied travel times; *Π* is the vector of route choice percentages as it were resulted by the route choice behavior of the travelers in correspondence to the real actual travel times (*C*). By indicating as *ξ* the percentage of travelers that have access to ATIS information (percentage of *equipped* travelers or ATIS *market penetration*), by assuming that a non-equipped traveler behaves with respect to route choice similarly as an equipped but not compliant one and, finally, by referring also to (1), it results that:

$$P(i/notEquipped) = P(i/NC) = \pi_i^N$$
$$\xi \cdot \eta \cdot \pi_i^Y = \lambda \cdot \pi_i^Y \quad (\xi \cdot \eta = \lambda) \tag{4}$$
$$[(1 - \xi) + \xi\,(1 - \eta)] \cdot \pi_i^N = (1 - \eta \cdot \xi) \cdot \pi_i^N = (1 - \lambda) \cdot \pi_i^N$$

Where the sum of $\lambda\,\pi_i^Y$ and $(1 - \lambda)\,\pi_i^N$ represents the contribution of the total demand on route *i*. In the following we assume that *ξ* is an exogenously given value (obtained by a direct estimate or by a market penetration model, for sake of simplicity but without lack of generality here not treated) and that the probability to be compliant surely depends on the (in)accuracy of the ATIS (however defined and computed) as in the following (5), that is expected to be a strictly decreasing function. Note that, for terminological simplicity, we will refer in the following *λ* as the *compliance*, even if it includes also the market penetration, which is, in practice, the maximum compliance (the one corresponding to the null inaccuracy).

$$\lambda = \lambda(Inacc) \tag{5}$$

3   Network Effects

All previous equations have been written with reference to a single origin/destination pair and with reference to a generic traveler. Now we assume that the network has more origin destination pairs, that for sake of simplicity the market penetration and the (in)accuracy are independent on origin/destination pairs and, finally, that on each O/D pair all the travelers behave as the general one referred in section 2. Moreover, we indicate as d ($n_{od} \times 1$ where n_{od} is the number of O/D pairs) the demand vector, as F ($n_r \times 1$ where n_r is the number of routes of the network) the vector of route flows, as f_{ATIS} ($n_a \times 1$ where n_a is the number of links) the vector of link flows, as p^Y ($n_r \times n_{od}$) the matrix of route choice probabilities for equipped and compliant travelers, as p^N ($n_r \times n_{od}$) the matrix of the route choice probabilities for not-equipped or not-compliant travelers.

$$
P^Y_{i,od} = \begin{cases} \pi^Y_i \text{ if the route i connects the od pair} \\ 0 \text{ otherwise} \end{cases}
$$
$$
P^N_{i,od} = \begin{cases} \pi^N_i \text{ if the route i connects the od pair} \\ 0 \text{ otherwise} \end{cases}
$$
(6)

Assuming at least at that the assignment model is based on a doubly steady state approach (absence of both day-to-day and within-day dynamics) and indicating by A the link/route incidence matrix ($n_a \times n_r$), the link flows are:

$$
f_{ATIS} = A \cdot F = A \cdot \left[\lambda \cdot p^Y (G) + (1 - \lambda) \cdot p^N (C) \right] \cdot d
$$
(7)

Where the compliance λ depends on the (in)accuracy of the ATIS (5) and the travel times G and C are, respectively, the one suggested by the ATIS and the actual ones. The actual costs depend, in turn, on the congestion mechanism via the dependency of link travel times (c) on link flows:

$$
C = A^T \cdot c(f_{ATIS})
$$
(8)

3.1   Recurrent and Equilibrated Conditions

In case of recurrent and equilibrated traffic conditions (7) and (8) hold without any further modification and two more sub-cases can occur: the ATIS is perfectly accurate (the compliance results equals the market penetration $\lambda = \lambda_{max} = \xi$); the ATIS is at some level inaccurate. In the first of the previous sub-cases the ATIS travel-times coincide, a part from random differences (treatable within their random distribution) with the actual travel times ($G{=}C$) and (7) and (8) can be solved by using a fixed-point approach.

$$
f^*_{ATIS} = A \left[\lambda_{max} p^Y \left(A^T c \left(f^*_{ATIS} \right) \right) + (1 - \lambda_{max}) p^N \left(A^T c \left(f^*_{ATIS} \right) \right) \right] d
$$
(9)

It is evident that the presence of ATIS impacts on traffic patterns only because equipped (and compliant) users could behave in different ways. It is typically hypothesized in these cases that the random dispersion of equipped and compliant users has a smaller variance than the one of non-equipped or non-compliant ones; otherwise ($p^Y = p^N = p$), the solution of the assignment problem is exactly as in case of absence of ATIS (10).

$$f_{ATIS}^* = f^* = A \cdot p \left(A^T \cdot c \left(f^* \right) \right) \cdot d, \quad p^Y = p^N = p \tag{10}$$

So, in case of perfectly accurate ATIS the differences between the traffic patterns in presence and absence of ATIS only depends on the differences between the *informed* and *not-informed* route-choice model. In the sub-case that ATIS is not accurate the compliance does not reach the market penetration and the link flows are likely to be more different from the ones of (10), this has recurrently induced some analyst (recently [7]) to argue that the difference can be controlled in order to induce *system-optimum* traffic patterns. With our notation:

$$Find \ G^* \Rightarrow \lambda^* = \lambda(Inacc(G^*, \lambda_{max} = \xi)) :$$
$$f_{ATIS}^* = f(\lambda^*, G^*, f_{ATIS}^*) = f_{SO} \neq f^* \tag{11}$$

In practice, such an attempt is likely to be unsuccessful, at least where system-optimum traffic patterns are different enough from user-optimum ones. This can be verified by using a simple two-link network; even if the example is really simple, there are not evidences that do not permit inductive extrapolation of the result. Consider a two-link network with a single O/D pair where the capacity of one of the links is assumed as a parameter to which, given the O/D demand, correspond different system-optimum traffic patterns. As a function of such a parametrical capacity different potential total travel time savings can be computed as the difference between user-optimum and system-optimum. In our network we have hypothesized that cost functions are BPRs and that the route-choice model is a simple multinomial-logit with systematic utilities equal to the opposite of route travel times. In order to move the traffic patterns from their natural user-optimum conditions a (descriptive, in our example) ATIS is introduced, which supplies fake travel times. However, to allow redirection of the required amount of flows, a minimum amount of compliance is required and it can be demonstrated, by using previous (7) and by imposing non-negativity constraints on route flows, that this can be computed as:

$$\lambda_{min} = max \left\{ 1 - \frac{f_{ATIS}^1}{d \cdot p_1^N}, 1 - \frac{d - f_{ATIS}^1}{d \cdot (1 - p_1^N)} \right\} \tag{12}$$

In previous equation f_{ATIS}^1 and p_1^N are the flow on one of the links of the network and the route choice probability for non equipped or non compliant travelers related to same link. According to (5), the actual compliance to ATIS information is not an independent variable; rather, it depends on the introduced inaccuracy which depends on the actual travel times (C - which are required to be the system-optimum ones) and on the ATIS travel times (G - which are known, given that they are the solution of the problem stated by (11). In the following Fig. 1

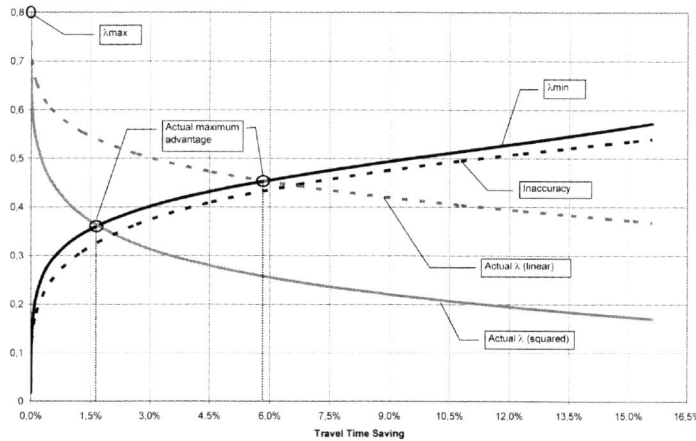

Fig. 1. Feasibility point for inducing system-optimum traffic patterns through ATIS

are plotted λ_{min}, actual λ (as a function of the inaccuracy) and the inaccuracy (the latest only for completeness) against the maximum potential advantage; feasible solutions are where the λ_{min} curve and the actual λ one intersect. In our case, even by considering an optimistic market penetration of 80%, the feasible point corresponds to a potential advantage of about 6% or 1.5% depending on if the actual lambda is considered to be approximable by a linear or a squared decreasing function of the inaccuracy. So, with reference to recurrent/equilibrated traffic conditions, the cost/opportunity of inducing system-optimum traffic patterns seems to be tipically low.

3.2   Recurrent and Equilibrated Conditions

In order to simulate the trajectories of the traffic network for non-steady-state conditions, further arrangements have to be introduced in previous (7) and (8); in particular, a dynamic process formulation (refer to [8]) has been introduced.

$$f^t_{ATIS} = \alpha A \left[\lambda^t p^Y(G^t) + (1 - \lambda^t)p^N(V^t) \right] d + (1 - \alpha)f^{t-1}_{ATIS}$$
$$V^t = (1 - \beta)V^{t-1} + \beta C^{t-1}$$
$$C^t = A^T c^t = A^T c(f^t_{ATIS}) \tag{13}$$
$$\lambda^t = (1 - \gamma)\lambda^{t-1} + \gamma\lambda(Inacc^t)$$

Where: all the variables already introduced have been referred to the generic day t of the deterministic dynamic process; V^t represents the route-choice systematic utility at day t; α represents the percentage of travelers that do reconsider the choices of the previous day; β and γ represent the convex weight of the actual travel times and of the compliance in their day-to-day updating process. What happens to the dynamic trajectories has been once again verified on a two-link network by using (13) and (5). Assuming some reasonable values for the dynamic

Table 1. Tested scenario

Case	Type of Information	Sub-case	Compliance	Market Penetration
1	Static / steady-state	a	Rigid	0.6
		b	Elastic	0.6
		c	Elastic	0.7
2	Dynamic / Accurate	a	Elastic	0.6
		b	Elastic	0.3

process parameters ($\gamma = 0.4$, $\alpha = 0.5$ and $\beta = 0.6$), the cases of Table 1 have been considered.

Three cases (1.a, 1.b and 1.c) assume that the ATIS supplies all days the same information (the route travel times that hold if a stochastic user-equilibrium were reached by the system) and: a) the compliance is considered to be rigid and equal to the value $\xi = \lambda_{max} = 0.6$; b) the compliance is a (more than linearly) decreasing function of the inaccuracy and the market penetration is 0.6; c) the compliance is as in case b) but its maximum value is 0.7. Two cases (2.a and 2.b) assume that the ATIS supplies all days accurate information (so that the compliance is identically equal to the market penetration of the ATIS) and: a) the market penetration is 0.6; b) the market penetration is 0.3. Figure 2 shows the result of experiment 1.a in terms of flow on one of the links, the dynamic process of the compliance is not shown, given that it is fixed; the dark line represents the system dynamics without ATIS, while the thin line refers to the presence of ATIS. Figures 3 and 4 show the result of experiment 1.b and 1.c, in these cases both the link flow and the compliance are shown.

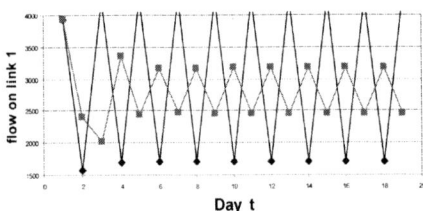

Fig. 2. Result of experiment 1.a ($\lambda_{max} = 0.6$)

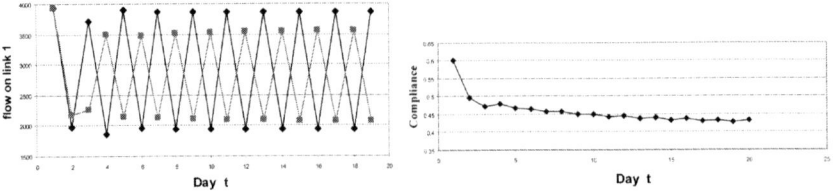

Fig. 3. Result of experiment 1.b ($\lambda_{max} = 0.6$)

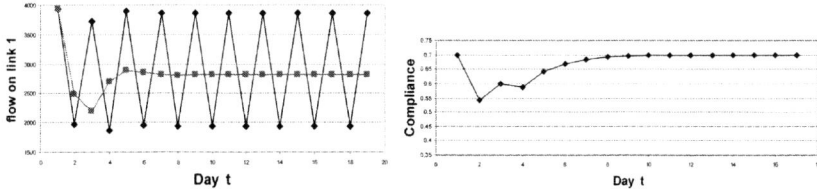

Fig. 4. Result of experiment 1.c ($\lambda_{max} = 0.7$)

Note that in cases 1.a and 1.b the system does not reach a stable configuration. In case 1.a (rigid compliance) the presence of ATIS seems to reduce
system's oscillations with respect to the absence of ATIS in a more effective way
than in case 1.b; however, case a) is unrealistic, provided that the compliance is
erroneously considered to be constant.

In case 1.c the system is able (due to the high value of the market penetration) to reach the stability and the compliance's dynamics shows that the higher
market penetration enables the whole system to self-correct the effects of the
initially inaccurate information strategy. Figure 5 shows the result of experiments 2.a and 2.b, the compliance are not shown because, in case of accurate
information, they are by definition equals to the market penetration (0.6 for 2.a
and 0.3 for 2.b).

Both in case 2.a and 2.b the system dynamics became stable and, due to the
intrinsic better performances of the accurate information strategy, this happens
both for a higher and a lower level of market penetration. Obviously, the lower
the market penetration the slower is the convergence. The previous numerical
results have suggested a more general and theoretical investigation of the system
dynamics stability in presence of ATIS. In the hypothesis of perfectly accurate
ATIS, the stability of the process has been investigated with reference to local
properties, provided that the theoretical investigation of global stability is not
affordable in general cases.

The result shown in fig. 6 have been proofed for general networks (for details
refer to [9]); the proof is based on the analysis of the eigenvalues of the dynamic
process described by (13) and (5), similarly to what has been made by [8].
The Argand-Gauss representation of the eigenvalues in fig. 6 shows that, in

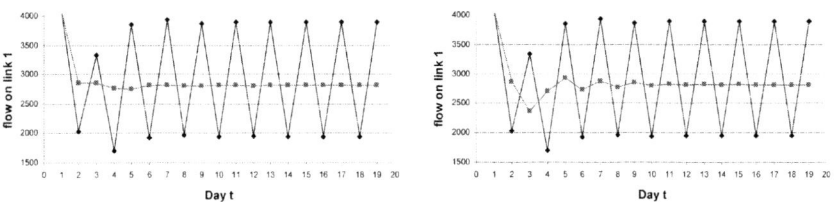

Fig. 5. Result of experiment 2a and 2.b ($\lambda_{max} = 0.6$) ($\lambda_{max} = 0.3$)

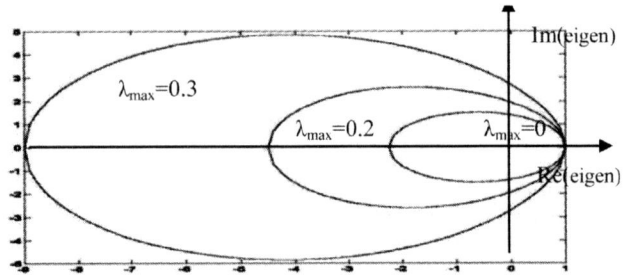

Fig. 6. Stability region for the deterministic dynamic process in presence of ATIS

case of accurate information and in case that the simulation model deals with explicit compliance elasticity, the stability region of the fixed point attractor monotonically increases with increasing market penetration. This is a desired result for ATIS applications and contradicts some common results that show instability phenomena in case of increasing ATIS market penetration. Such a not negligible difference is due, to our opinion, to the lack of other models in explicitly simulating compliance and to the fact that most of the previous models does not deal with accurate information.

4   Conclusions

It is crucial to properly deal with extra-uncertainty that ATIS can induce, in particular with the relashionship between travellers' compliance and ATIS inaccuracy. Missing in modelling uncertainty and compliance leads to incorrect assessment of ATIS effects and/or unsuitable objectives, such as the one of using ATIS to induce system-optimum conditions. The best advantage of ATIS seems to be in increasing traffic stability and improve transition between different recurrent conditions. Moreover, ATIS can be crucial for mitigating network impacts of non-recurrent conditions and it is expected that the compliance that can be spent is the one gained in recurrent conditions.

References

1. Avineri, E., Prashker, J.N.: The impact of Travel Time Information on Traveler's Learning under uncertatinty. Transportation 33(4), 393–408 (2006)
2. Chen, P., Srinivasan, K., Mahmassani, H.S.: Effect of Information Quality on Compliance Behaviour of Commuters under Real-time Information. Transportation Research Record 1676, 53–60 (1999)
3. Bottom, J., Ben-Akiva, M., Bierlaire, M., Chabini, I., Koutsopoulos, H.N.: Investigation of Route Guidance Generation Issues by Simulation with DynaMIT. In: 14th International Symposium on Transportation and Traffic Theory, Jerusalem (1999)
4. Bierlaire, M., Crittin, F.: New algorithmic approaches for the anticipatory route guidance generation problem. In: 1st Swiss Transportation Research Conference (2001)

5. Henn, V., Ottomanelli, M.: Handling uncertainty in route choice models: from probabilistic to possibilistic approaches. European Journal of Operational Research 175 (2006)
6. Dell'Orco, M., Kikuchi, S.: An alternative approach for choice models in transportation: use of possibility theory for comparison of utilities. Yugoslav Journal of Operations Research 14(1) (2004)
7. Yu, J.W., Peeta, S.: A Behavior-based Network Modeling Approach for Deployable Information Provision Strategies. In: TRB 2007 Annual Meeting Compendium of papers, CD-ROM (2007)
8. Cantarella, G.E., Cascetta, E.: Dynamic processes and equilibrium in transportation networks: toward a unifying theory. Transportation Science 29, 305–329 (1995)
9. Bifulco, G.N., Simonelli, F.: The effect of ATIS on transportation systems: theoretical analysis and numerical applications. In: XI Urban Transport and the Environment for the 21st century, pp. 723–732 (2005)

Human-Like Adaptive Cruise Control Systems through a Learning Machine Approach

Fulvio Simonelli, Gennaro Nicola Bifulco, Valerio De Martinis, and Vincenzo Punzo

University of Napoli Federico II, Department of Transportation Engineering Via Claudio 21, 80125 Napoli, Italy
{fulsimon,gennaro.bifulco,vdemartinis,vinpunzo}@unina.it

Summary. In this work an Adaptive Cruise Control (ACC) model, with human-like driving capabilities,based on a learning machine approach, is proposed. The system is based on a neural network approach and is intended to assist the drivers in safe car-following conditions. The proposed approach allows for an extreme flexibility of the ACC that can be continuously trained by drivers in order to accommodate their actual driving preferences as these changes among drivers and over time. The model has been calibrated against accurate experimental data consisting in trajectories of vehicle platoons gathered on urban roads. Its performances have been compared with those of a conventional car-following model.

Keywords: Advanced Driving Assistance Systems (ADAS), Adaptive Cruise Control (ACC), Car Following, Neural Networks.

1 Introduction

In last years Advanced Driving-Assistance Systems (ADAS) have received an increasing attention. A broad picture of many possible applications and objectives in the field of automation, control and assistance in driving choices and actions can be found in [1]. It is worth noting that, among all applications, we are not here interested neither in AHS (Advanced Highway Systems) nor in applications aimed at maximizing traffic throughputs (for istance [2]). Rather, we will focus on individual-vehicle not-cooperative applications, mainly oriented to increase driving comfort. Among ADAS Collision-Warning/avoidance System (CWS) and Advanced/Adaptive Cruise Control (ACC) systems play an important role in ADAS field also for the automotive industry. Both CWS and ACC are aimed to increase driving safety: CWS directly by avoiding crashes and ACC (which generally include the previous) also by improving user comfort through reduction of the mental and physical workload [3]. Several research and industrial projects (e.g. [4]) have been recently oriented to develop ACC systems that can warn the driver and/or automatically adjust the vehicles' speed. This is mainly achieved by using forward-looking radar to detect speed and distance of the vehicle ahead and to suggest and/or automatically adjust speed in order to ensure safe speed. It should be noted that these systems must earn the

E. Avineri et al. (Eds.): Applications of Soft Computing, ASC 52, pp. 240–249.
springerlink.com   © Springer-Verlag Berlin Heidelberg 2009

drivers' compliance in order to be actually used and effective; such an issue has oriented the researchers toward human-like ACC. This means that ACC cannot be based only on cinematic and physical considerations strictly ensuring safety headways and/or speeds, but they have to imitate individual driver behaviours in order to achieve his/her maximum comfort and compliance to the system. In other words not only has to be ensured absolute safety, (namely the headway and/or speed that must not be overcome) but also applied the most accepted driving behaviours among the several ones not exceeding the absolute safety thresholds. Therefore the challenge is also in reproducing individual driver trajectories/behaviours. Reproducing driving trajectories has been one of the main aim of modelling longitudinal motion of vehicles in a traffic flow; this kind of models, also known as car-following models, have been traditionally developed as one of the main sub-models for the microscopic simulation of traffic flows; they also represent a theoretical reference for many ACC systems. In car-following literature several classes of models are generally recognized (for a detailed review see for ex. [5]). The most known are: a) stimulus-response models (e.g. GHR model [6]); b) safety distance or collision avoidance models (e.g. Gipps' model [7]) ; c) psychophysical or action-point models (e.g. [8]). As mentioned, these models have recently gained attention also in defining normative behaviours and their adoption in ACC system. In particular, due to its mathematical properties guarantying an indefinite stable (safe) behaviour, the Gipps' model has been widely applied in microsimulation packages and it will be considered in the following for reference. By trying to calibrate a car-following model against accurate experimental data, some authors e.g. [9] [10] [11] have verified that models' parameters should be considered as dependent on different actual conditions (such as the attitude and skill of the driver) and can vary, for a given driver, depending on the reason and duration of the trip or on other parameters specific of each trip or of the roads crossed in the trip. In other terms, safe driving behaviours, still respecting absolute safety constraints, may change in different context and/or for different drivers. It has been also argued that driving is in fact an adaptive process [12]; the behavior and the physical parameters describing it change over time and space. The parameters of any car-following model should then sensibly vary along a trip, even for the same driver: not only average values but parameters' trajectories should be therefore estimated. The previous considerations are gaining increasing attention in the field of simulation-oriented car-following models and these seems to have even greater relevance if car-following models are used in ACC systems: an adaptive car-following model is of crucial importance in order to reproduce human-like driving behaviour. It is therefore valuable to investigate the possibility of generating an adaptive system capable of matching drivers' attitude coupled with a crash avoidance system that prevents violations of absolute safety. However, this is only partially the approach that can be observed even in recent research projects [13] [14] where the parameters of the control logic are pre-set when the system is produced. These are calibrated against generic observed vehicle trajectories that do not reflect the driving style of the driver that will actually use the system nor the context in which he/she will drive. In this sense, such an approach does not ensure a fully adaptive system. It seems to the authors that the ACC systems so far developed are intrinsically *quasi human-like* and that their reference-human is a general one and not an actual person. The point of view proposed in this paper is different: we propose a preliminary study on the possibility of building a learning machine, in particular based on a neural network approach, capable of learning (continuously and/or on request, in order to fulfil adaptive issues) the driving attitude of the actual driver in a given trip or part of a trip. In other words, when the control of the vehicle is in charge of the driver, the system learns his/her behaviour. In this case, whenever the driver is not satisfied by the automatic controller, he/she can train it again by driving for a few minutes more. This would be far more appealing on the market. To develop such a tool is necessary to understand what the relevant parameters affecting driving are. To this aims, also in this paper we will start from the studies that have been carried out in the area of car-following models and we will also compare our model to the performances of a simulation oriented car-following model. However, it has to be clear that the point of view from which we look at the phenomenon and the justification of the study are different and allow for non-conventional use and comparison of the reference car-following model. In the following of this paper, section 2 will present the experimental data on which the proposed approach have been developed, section 3 will describe the proposed model and the experiment designed to calibrate and validate it.

2   Experimental Data

Data used in this study have been extracted from a series of experiments carried out along roads in areas surrounding of Naples (Italy), in conditions of real traffic, between October 2002 and July 2003. Experiments were performed by driving four vehicles in a platoon along urban and extra-urban roads, in different traffic conditions. Vehicles were all equipped with Kinematic differential GPS (K-dGPS) receivers recording the position of each vehicle at intervals of 0.1 seconds. Only one-lane-per-direction roads were considered, such that car-following behaviour was unaffected by other behaviour like lane-changing. Data collected via equipped vehicles in real contexts have characteristics of high realism, as needed by our experiment; however, as frequently found in literature, they are relatively limited with respect to length due to technical constraints especially in urban environments. Three trajectory data sets have been extracted from the data base for this work, respectively named 30A, 30B and 30C. Sets 30A and 30C are relative to one-lane urban roads while 30B is a two-lane (one for direction) rural highway that by-passes the historical centre of Pozzuoli (a town near Naples). Driving patterns are qualitatively different among routes: urban patterns show repeated vehicle stops, the other being more regular. From GPS positional data, time series of inter-vehicle spacing are immediately available while vehicle speeds and accelerations have to be calculated through successive derivations of the space travelled. In spite of the expected precisions of dGPS

Table 1. Experimental data: VL = Leader Speed (m/s), Δx = Spacing (m), ΔV = ΔSpeed (m/s)

	Trajectories 1-2			Trajectories 2-3			Trajectories 3-4		
Experiment 30A; Length: 1354.4 meters (188.8 seconds)									
	VL	Δx	ΔV	VL	Δx	ΔV	VL	Δx	ΔV
max	14.27	17.52	2.88	14.66	22.29	3.47	14.47	20.12	2.57
min	0.00	4.33	-2.89	0.00	4.28	-2.42	0.00	4.34	0.00
mean	7.34	10.96	0.03	7.31	12.71	0.07	7.24	11.98	0.06
var	19.64	14.60	0.70	19.88	27.87	0.71	18.89	21.68	0.58
Experiment 30B; Length: 2486.5 meters (247.8 seconds)									
	VL	Δx	ΔV	VL	Δx	ΔV	VL	Δx	ΔV
max	18.58	20.12	2.68	18.37	31.80	2.22	18.95	17.97	2.18
min	0.00	5.05	-3.08	0.00	4.76	-2.84	0.00	4.92	-3.17
mean	9.99	11.53	-0.01	10.00	15.48	-0.02	10.01	11.71	-0.02
var	25.02	11.37	0.55	24.98	44.67	0.71	25.01	10.47	0.68
Experiment 30C; Length:2183.2 meters (349.5 seconds))									
	VL	Δx	ΔV	VL	Δx	ΔV	VL	Δx	ΔV
max	14.64	30.94	3.92	14.13	27.78	5.04	14.11	19.61	2.78
min	0.00	3.93	-6.75	0.00	4.57	-5.69	0.00	4.10	-3.20
mean	6.25	9.09	0.01	6.94	11.29	0.00	6.24	8.53	-0.01
var	20.50	20.22	1.11	20.27	31.40	1.21	20.00	13.38	0.68

positional measurements, data needed to be filtered in view of the high measurement noise due to the urban environment on the one hand (e.g. multi-path effect) and to stringent requirements of this work. The core of the problem was to filter noisy trajectory data of each vehicle without altering platoon data consistency, i.e. speeds and accelerations of following vehicles had to be estimated so that inter-vehicle spacing calculated from them were equal to the real one. Otherwise, for example, even slight differences between estimates and actual speeds of a vehicle could easily have entailed negative spacing in case of a stop. This was accomplished by means of a non-stationary Kalman filter. Differently from other experiences aimed only at the smoothing of speed and acceleration profiles of individual vehicles (see e.g. [15]) the designed filter was conceived and applied to the whole platoon considering following vehicles as a sole dynamic system and allowing to solve one consistent estimation problem instead of several independent (and inconsistent) ones (see [16] for the details). Thus accurate and consistent time series data of inter-vehicle spacing, speeds and accelerations were obtained. Since the focus of the work is on vehicle interactions, trajectory data are referred to the couples of vehicles rather than to each single vehicle. The three couples are named 1-2, 2-3, 3-4 where the numbers refer to the order in the platoon. The main characteristics of the obtained trajectories have been reported in Table 1, as evident in Table 1, the driving conditions to which the data are related include stop-and-go and trajectories refer to congested traffic conditions.

3 The Proposed Model

The possibility of using an ANN is investigated in this work as opposed to using
a traditional car-following model for learning and reproducing the behaviour of
a given driver; this has to be framed in the field of ADAS applications, rather
than on the one of microscopic traffic simulation applications. Further to the
advantages deriving from increased flexibility, a model based on a neural network is also suitable in terms of learning speed whoch is crucial when real time
learning is required. A neural network model requires the identification of the
input-output (I/O) to be reproduced and the definition of the type and architecture of the network to be used. The reference I/O has been chosen by taking
cue from car-following models. In particular, analogously to the Gipps model,
inputs are the spatial spacing between leader and follower and the speed of the
follower and the output is the speed planned by the follower for the following time step. Two different kinds of ANN have been trained. A first type is a
feed-forward network with one hidden layer characterized by a sigmoid transfer function. It has been chosen being the most consolidated, frequently used
and relatively easy. Several structures have been tested for the network and the
chosen one is made of 10 neurons in the hidden layer. The training was carried
out through a back-propagation technique using the Levenberg-Marquardt algorithm. Moreover, the neural network has been trained several times according to
exogenously reasonable fixed values of the time offset between input and output
variables, that in car-following stands for the time delay of the response. The
variation range of those offsets has been chosen on the basis of the variation
range of the reaction times resulting from car following models estimation results available in the literature. In more detail, 4 offset values (0.4, 0.6, 0.8 and
1 second) have been taken into account. The more appropriate resulting offset
has been 0.6 seconds. A second type of ANN is a recurrent network, operating
as a sequential system associating output patterns to input patterns in accordance with a dependence on an internal state, which in turn evolves on the basis
of the inputs. In particular it is used an Elman network; this choice is due to
the fact that it introduces a feedback connection from the output of the hidden
layer to its input, so as to be able to recognize dynamic patterns (that are expected to hold in car-following phenomena). Also in this case the hidden layer
comprises 10 neurons and the training set is the same as for the feed-forward
networks. Obviously, in this case the input-output pairs should be presented to
the network as an ordered sequence, consistent with the dynamics of the experimental trajectories. As term of comparison for the evaluation of the training
performances of both neural networks, a traditional car following model (Gipps
model) has been also calibrated on the same training set, trying to minimize the
error in the input-output relationship, namely looking for the set of parameters
leading to a model output (i.e. the value of speed after the reaction time lag) as
close as possible to the experimental one. This kind of estimation differs from
the usual trajectory-based calibrations performed in previous works e.g. [9] [10]
[11], where system dynamics were properly taken into account. In this paper
the Gipps model has been calibrated as if it was a static I/O relationship. In

fact the not necessarily ordered pairs of I/O experimental data have been used as inputs to the calibration process instead of the leader's and the follower's trajectories used in [9] [10] [11]. It is worth mentioning that such a type of I/O estimation of the Gipps' model is quite unfair but requires a computational time similar to that required for the neural networks training (from 5 to 30 seconds in our trials depending on the length of the trajectory), while the more appropriate trajectory-based estimation requires significantly larger computational times (several minutes) making it not suitable for real-time learning contexts.

3.1   Calibration and Validation of the Proposed Model

Training/calibration results, coming from the experiments previously described, are shown in Table 2, where the training errors are expressed in terms of mean absolute percentage error (MAPE) and root mean square error (RMSE). The results are shown for the two types of ANN and for the Gipps' model described in the previous sections. The ranking in terms of (decreasing) ability of fitting data is: Elman ANN, FFNN, Gipps. Indeed the Elman ANN have fitted better (with the exception of case 3-4 of experiment 30B, where it fits practically as the FFNN) the data for its own nature: that is the one of being a model with memory. And this is also the reason why it has to be necessarily trained with time series of data instead of mere I/O data sets. As already mentioned Gipps' model has been used as if it was a static model and calibrated on I/O data sets; naturally having the Gipps' relationship less degrees of freedom than an ANN it results a lower ability of fitting experimental data.

The capability of input-output trained ANNs in reproducing follower's behaviour when used as car-following models has been checked through simulation of follower's trajectory. To verify the behaviour of the follower as it is simulated by the models, the calibrated relationships are fed with the leader's experimental trajectory. The simulated time series of follower's speed and inter-vehicle spacing are then compared with the experimental ones by means of MAPE (Mean Absolute Percentage Error) and RMSE (Root Mean Square Error) statistics. This test is a verification of the (calibrated) relationships' ability in reproducing

Table 2. Calibration results

		Feed Forward		Elman		Gipps	
Experiment	Trajectory	MAPE	RMSE	MAPE	RMSE	MAPE	RMSE
30A	1-2	2.98%	0.156	2.76%	0.150	8.63%	0.346
	2-3	3.14%	0.149	2.30%	0.100	8.28%	0.328
	3-4	2.76%	0.178	2.85%	0.146	9.59%	0.520
30B	1-2	1.45%	0.151	1.30%	0.109	5.50%	0.390
	2-3	1.70%	0.144	1.39%	0.090	6.66%	0.460
	3-4	2.69%	0.211	2.55%	0.170	6.79%	0.450
30C	1-2	6.87%	0.218	4.45%	0.130	10.23%	0.290
	2-3	4.08%	0.191	3.57%	0.140	12.21%	0.360
	3-4	4.75%	0.194	3.82%	0.130	11.12%	0.330

Table 3. Simulation test (speeds)

Experiment	Trajectory	Feed Forward MAPE	RMSE	Elman MAPE	RMSE	Gipps MAPE	RMSE
30A	1-2	11.07%	0.706	11.27%	0.530	13.95%	0.656
	2-3	11.53%	0.582	8.56%	0.410	11.99%	0.533
	3-4	12.03%	0.552	9.25%	0.460	11.85%	0.560
30B	1-2	7.77%	0.738	7.63%	0.699	14.39%	1.030
	2-3	8.65%	0.688	6.62%	0.550	12.67%	0.887
	3-4	10.24%	0.918	7.69%	0.654	16.49%	1.081
30C	1-2	15.85%	0.730	15.92%	0.661	28.06%	0.886
	2-3	14.47%	0.693	15.43%	0.680	19.91%	0.688
	3-4	14.76%	0.867	16.58%	0.692	18.01%	0.942

the follower's behaviour, i.e. when they are used as dynamic models. It is worth nothing that, as discussed in the introduction, the models are intended to reproduce a human-like driving behaviour and that absolute safety issues are, in our approach, of pertinence of a coupled collision avoidance system; from this point of view, the results should be read in terms of aggregate capability of reproducing the drivers' behaviours and errors in terms of spacing should not generate in the readers warnings about safety. Looking at both MAPE and RMSE indicators for speed results (Table 3), the overall consideration can be made that Elman and FFNN perform similarly (the previous slightly better that the FFNN) and better than Gipps. These results seem quite straightforward from results of training. In particular, the poor performances of Gipps can be justified by the same considerations already done. The results about spacing (Table 4) are particular meaningful since this variable has not been used to train the relationships and for this reason is an indicator of the ability of the calibrated I-O relationships of capturing the system dynamics. If we look at the error statistics on spacing we observe that the Elman ANN perform almost always (8/9) better than the FFNN, that in turn outperforms the Gipps' model. The fact that the Elman

Table 4. Simulation test (spacing)

Experiment	Trajectory	Feed Forward MAPE	RMSE	Elman MAPE	RMSE	Gipps MAPE	RMSE
30A	1-2	8.25%	1.399	6.98%	0.996	19.35%	2.631
	2-3	8.71%	1.471	6.54%	0.898	14.14%	2.213
	3-4	8.90%	1.489	8.25%	1.285	13.46%	0.1.911
30B	1-2	17.98%	2.930	12.99%	1.816	55.88%	9.701
	2-3	9.28%	1.996	8.88%	1.812	24.43%	4.963
	3-4	23.30%	3.647	12.99%	2.064	53.77%	11.01
30C	1-2	19.97%	2.939	16.15%	2.652	25.18%	4.087
	2-3	16.77%	2.717	16.38%	2.430	19.34%	2.800
	3-4	19.62%	2.221	19.76%	2.628	22.38%	6.865

Table 5. Validation

Experiment		Feed Forward		Elman		Gipps	
30C	Trajectory	MAPE	RMSE	MAPE	RMSE	MAPE	RMSE
Speed	1-2	13.75%	0.897	26.77%	1.320	43.46%	1.123
	2-3	21.19%	1.013	22.78%	0.950	27.20%	0.933
	3-4	11.40%	0.804	17.89%	1.364	18.92%	1.065
Spacing	1-2	16.44%	2.038	20.29%	3.381	34.16%	5.048
	2-3	28.37%	5.997	28.94%	6.071	24.61%	4.479
	3-4	22.98%	3.333	29.68%	4.444	5.719%	9.307

ANN has performances similar to the FFNN in terms of speeds (5/9) while better performances (8/9) in terms of spacing, is not a surprise being the model
dynamical.

Finally, in order to test the suitability of the models to reproduce the behaviour of a given driver, they have been validated (Table 5 and Figure 1) by
training on 60% of the trajectory and simulating the rest of the trajectory. This
kind of validation, making use, for each of the trajectories, of the data of the
related driver is consistent with our applicative context, which does not require
generalization among drivers. Obviously, in order to have an effective training
we need an enough large training set and for this reason the validation test has
been carried out only for the longest trajectory (30C).

As expected, the Gipps' model has very poor performances in validation.
This is surely due to the fact that the I/O approach in which has been here
framed is strongly inadequate. The Elman ANN performs worst than the FFNN
in validation: the authors opinion is that this happens because for the Elman
ANN it is much more likely to incur in over-fitting problems. Figure 1 shows
the comparison among the experimental trajectory (in terms of follower speed)
and the modeled one, they are referred to the FFNN model and to trajectory
2-3 of experiment 30C (which is the trajectory that performs worst in Table 5).
Both the simulation part of the experiment and the validation part are shown.
As evident from cumulated frequencies of relative errors, about 70% (more than
75% in case of simulation) of the errors are less or equal to 10%, moreover the

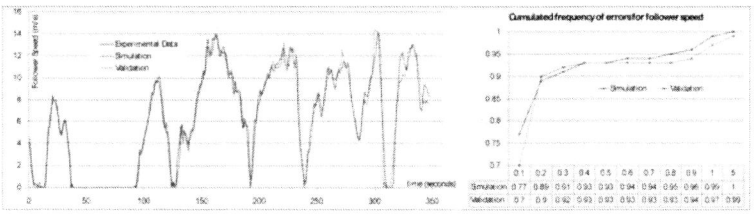

Fig. 1. Simulation and validation result for trajectory 2-3 of experiment 30C (FFNN
model)

cumulated frequencies for validation are similar the one of the simulation that indicates an adequate robustness. It is worth nothing that, in particular with reference to the FFNN, an intrinsically static model calibrated through a (static) I/O approach seems to be quite successfully applicable in a typically dynamic context (as it is the car-following one) or, at least, there are not evidences that such an approach cannot be used for our purposes.

4 Conclusions

In this paper it has shown the feasibility of a fully adaptive approach to ACC (Adaptive Cruise Control systems) based on a learning machine that can be quickly trained in real-time. The training time for the FFNN is terms of few seconds for data collected on trajectories of many seconds; this means that the on-line use is suitable. The system allows for reproducing quite well a human-like driving behaviour and the test performed on the proposed model suggest that it can be used for effective applications. The results only give feasibility indications; the work has to be continued in order to refine the approach and to validate it against more experimental data. Moreover, prototypal implementation of the system and/or further studies and simulations with more emphasis on practical implementation aspects could clarify if those kind of systems can have efficiency and robustness characteristics also with reference to contexts in which frequent human commands are given to (re)train the system. In conclusion, it seems to the authors that the obtained preliminary results could suggest new original analyses also on traffic simulation oriented car-following models.

References

1. Vahidi, A., Askandarian, A.: Research advances in Intelligent Collision Avoidance and Adaptive Cruise Control. IEEE Trans. Intelligent Transp. Systems 4(3), 143–153 (2003)
2. Ho, F., Ioannu, P.: Traffic flow modeling and control using Artificial Neural Networks. IEEE Control Systems 16(5), 16–26 (1996)
3. Tadaka, Y., Shimoyama, O.: Evaluation of driving-assistance systems based on drivers' workload. In: Proceedings of the International Driving Symposium on Human Factors in Driver Assessment, Training and Vehicle Design (2004)
4. VTT.: Deliverable 2/B of the Humanist Network of excellence: an inventory of available ADAS and similar technologies according to their safety potential (2006)
5. Brackstone, M., McDonald, M.: Car-following: a historical review. Transportation Research F 2, 181–196 (1999)
6. Gazis, D.C., Herman, R., Rothery, R.W.: Non-linear follow-the-leader models of traffic flow. Operations Research 9(4), 545–567 (1961)
7. Gipps, P.G.: A behavioural car-following model for computer simulation. Transportation Research B 15(2), 105–111 (1981)
8. Brackstone, M., Sultan, B., McDonald, M.: Motorway driver behavior: studies on car following. Transportation Research F 5(1), 31–46 (2002)
9. Brockfeld, E., Kühne, R.D., Wagner, P.: Calibration and validation of microscopic traffic flow models. Transportation Research Record 1934 (2004)

10. Ranjitkar, P., Nakatsuji, T., Asano, M.: Performance Evaluation of Microscopic Traffic Flow Models Using Test Track Data. Transportation Research Record 1876 (2004)
11. Punzo, V., Simonelli, F.: Analysis and comparison of car-following models using real traffic microscopic data. Transportation Research Record 1934, 42–54 (2005)
12. Hoogendoorn, S.P., Ossen, S., Screuder, M.: Adaptive Car-Following Behavior Identification by Unscented Particle Filtering. In: 86th TRB annual meeting (2007)
13. Alonso, M., Garayo, P., Herran, L.: PREVENT-SASPACE: Deliverable D20.33 - Functional requirements (2005)
14. Tango, F., Saroldi, A., Alonso, M., Oyarde, A.: Towards a new approach in supporting drivers functions. In: Proceedings of the ITSC 2005 8th International IEEE Conference (2005)
15. Ma, X., Andreasson, I.: Dynamic car-following data collection and noise cancellation based on Kalman smoothing. In: IEEE International Conference on Vehicular Electronics and Safety (2005)
16. Punzo, V., Formisano, D.J., Torrieri, V.: Nonstationary Kalman Filter for Estimation of Accurate and Consistent Car-Following Data. Transportation Research Record 1934, 3–12 (2005)

Fuzzy Evaluation of Contra-Flow Evacuation Plans in Texas Medical Center

Fengxiang Qiao, Ruixin Ge, and Lei Yu

Department of Transportation Studies, Texas Southern University, 3100 Cleburne Avenue,
Houston, Texas, U.S.A.
Tel.: (713) 313-1915; Fax: (713) 313-1856
{qiao_fg, ger, yu_lx@tsu.edu
http://transportation.tsu.edu

Abstract. With the needs of emergency preparedness due to various disasters such as Hurricane, flooding, radiological accident, terrorist attack, toxic material leakage, etc., an efficient transportation evacuation plan such as contra-flow is very crucial. Simulation is a superior tool that can measure the effects of different plans, however often offers contradictory Measurement of Effectiveness (MOE), which may not directly yield out an optimal evacuation plan. In this paper, Fuzzy Logic is employed to evaluate the evacuation plan. A case study illustrates the entire evacuation process that evacuates contra-flow evacuation plans in a typical multi-institutional area: Texas Medical Center in Houston, USA.

Keywords: Fuzzy Logic, Transportation Evacuation, Contra-Flow, Microscopic Simulation, VISSIM.

1 Background

In 2005, Hurricanes Katrina and Rita hit the USA Gulf Coast, resulting in mass evacuations with huge time delay and property loss on highways [1], [2]. A gunman killed in 2007 at least 30 people in one of two shootings on the campus of Virginia Tech [3], which awoke universities and institutions of paying more attentions on campus safety and evacuation. These are two extreme causes that need emergency preparedness [4]. Others may include flooding, radiological accident [5], terrorist attack [6], toxic material leakage [6], etc. As emergency situations generally involve seriously life and property damage, one good way to deal with it is to evacuate people and valuable materials from the affected area in a quickest way. Thus, an efficient transportation evacuation plan is very crucial and involves careful planning.

According to National Response Plan (NRP), an evacuation is "an organized, phased, and supervised withdrawal, dispersal, or removal of civilians from dangerous or potentially dangerous areas, and includes their reception and care in safe areas", and "the U.S. highway infrastructure permits the movement of large numbers of people over significant distances in a timely and safe manner to suitable shelters away from the hazard zone," [7]. As the travel demand when evacuating is much higher than in normal peak hour, special countermeasures should be taken into consideration

E. Avineri et al. (Eds.): Applications of Soft Computing, ASC 52, pp. 250–259.
springerlink.com   © Springer-Verlag Berlin Heidelberg 2009

such as contra-flow lane reversal, which is a program designed for quick emergency evacuation of an area, and targets on maximizing the capacity of transportation infrastructure and improving the efficiency of traffic operations [8].

One of the vivid and efficient ways to evaluate different evacuation plans including the contra-flow plans are through microscopic computer simulation. Microscopic traffic simulation analyzes flow of individual vehicles in response to driver and vehicle characteristics as well as traffic control devices. It has been increasingly used to model traffic flow under evacuation conditions to help evacuation operation/plan development. However, evaluating an evacuation plan is now and then hard since the generated Measurement of Effectiveness (MOE) can be contradictory. Moreover, an improvement in part of the network may result in performance decrease in another part, or even in the entire network.

Fuzzy logic provides a feasible way to evaluate evacuation plans. Tiglioglu used fuzzy logic procedure to determine the households' risk assessment and evacuation decision-making process according to the location where people live in and severity of the storm, [9]. Poudel et al applied fuzzy logic approach for aircraft evacuation modeling, [10]. Li used fuzzy synthetic assessment on safety evacuation measures in library, [11]. In this following part of the paper, fuzzy logic is employed to evaluate contra-flow evacuation in a typical dense Multi-Institutional Center: Texas Medical Center (TMC) in Houston, USA.

2   Simulating Contra-Flow Evacuation Plans

Texas Medical Center (TMC) in Houston, Texas has a dense population with 45 institutions as well as an abundant history of emergency events. There are seven major traffic generators (i.e. seven garages) within TMC area. The proposed evacuation plan(s) should be able to efficiently evacuate all vehicles inside that area.

The microscopic traffic simulation software VISSIM [12] is utilized as a simulation tool, which is a microscopic, time step and behavior based simulation model analyzing the full range of functionally classified roadways and traffic operations. It can address specific geometric designs of any transportation network, and provides Measurement of Effectiveness (MOEs) for each predefined operational scenarios with microscopic driving behaviors captured by its internal algorithm.

The general framework for the simulation of evacuation plans includes six major steps: incident assumption; field data collection; network coding; scenario development; parameter calibration; and traffic simulation. Figure 1 is the illustration of the study area in Texas Medical Center.

2.1   Incident Assumption

For the particular case in TMC, the fictitious incident is a truck crash causing hazardous material leakage inside the study area on a typical normal weekday. All persons have to evacuate onto outside the TMC area to avoid any possible poisonings. Figure 1 is the illustration of the study area in Texas Medical Center.

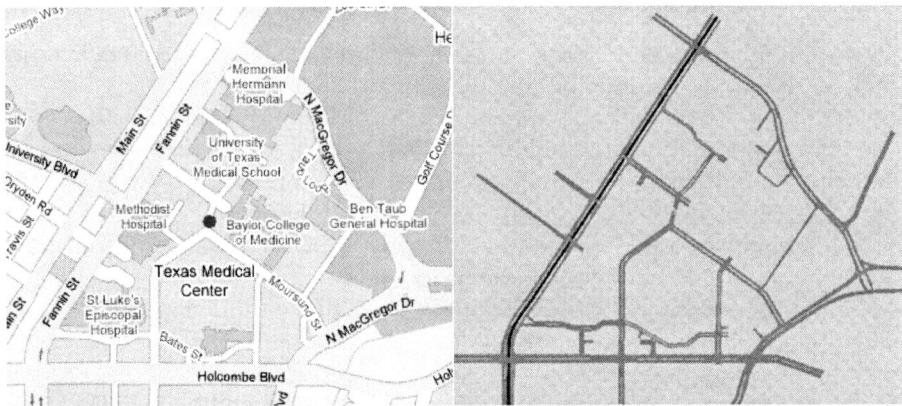

Fig. 1. Study Area: Texas Medical Center, map from google.com (left) and the network coded in VISSIM (right). The dot in the center of the graph shows the location of a fictitious incident related to toxic material leakage.

2.2 Data Collection

Field data collection is necessary since geometry design, locations of traffic control devices will be used to establish the sketch of road network for simulation, while real time traffic volume, signal timing plan and vehicle composition, etc. are indispensable for the calibration of the internal models of simulation software VISSIM. Global Positioning System (GPS) is a useful tool to collect instantaneous speed data.

2.3 Network Coding

The process is to develop the transportation network in study area, as well as prepare initial traffic parameters in the microscopic simulation model VISSIM. Traffic control devices (stop signs, yield signs, reduce speed areas and signals), vehicle information and route assignment setting are inputted to the simulation model. Data collection points can be set up in the simulation network, so as to reports the necessary operational traffic information.

Table 1.

Scenario	Countermeasures
1	Normal situation with no any assumed incident happened
2	Incident occurs, evacuation routes are assigned for evacuating vehicles with no any contra-flow strategy applied
3	Incident occurs, contra-flow at John Ave between garage 5 (G5) and Holcombe Boulevard, south bound
4	Incident occurs, contra-flow at Mousund Street between John Avenue and South MacGregor Drive, west bound
5	Incident occurs, contra-flow at Holcombe Boulevard between G2(E) and Braeswood Boulevard west bound
6	Incident occurs, contra-flow at Holcombe Boulevard between John Ave and Bertner Avenue east bound

2.4   Scenario Development

The designed evacuation scenarios are specified in Table 1. There are total 6 scenarios including no incident and do-nothing actions with incident. The garages, possible contra-flow links and most dangerous area when toxic leaking occurs are illustrated in Figure 1.

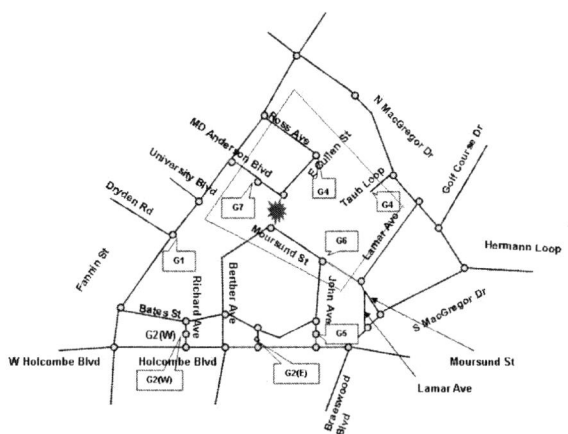

Fig. 2. Garages inside the Texas Medical Center (numbered after the letter "G"). The red links are where contra-flows are possibly applied. The blue frame area is called a 'core area' where toxic leaking has the most impact in.

2.5   Calibration

Field speed and simulation generated speed are compared each other to make sure the simulation results are consistent with the field conditions. Driving behavior parameters in VISSIM are calibrated by using computer program which is developed by adopting Genetic Algorithm as its core component.

2.6   Simulation

Simulation is conducted for 3600 seconds. Table 2 shows the whole network wide simulation results, where S1, …, S6 are the abbreviations of the six proposed scenarios, while Table 3 shows the core area performance through simulation.

Since scenario 1 is the normal time without incident and Scenario 2 is evacuation time without any improvement, their performances in Table 2 are reasonably in the two extreme ends. The percentage of vehicle left the network in Scenario 3 is 3.6% less than Scenario 4. Yet, other measure of effectiveness (MOE) such as average travel time, average speed, average delay and average number of stops all indicate that contra-flow plan in Scenario 4 is better than Scenario 3. Also, it seems that contra-flow on Holcombe Blvd heading west is better than the one heading east. However in Table 3, we can see heading west result in a worse performance in the core area which is most affected by toxic material leakage. This is why a smart evaluation methodology such as Fuzzy Logic based on is to be employed to deal with these contradictory MOEs.

Table 2. Network Wide Performances through Simulation under Six Proposed Scenarios

	S1	S2	S3	S4	S5	S6
# vehicles left network	8390	4602	4602	4964	6144	5920
# vehicles left garages	8876	5002	5018	5436	6406	6208
% vehicles left network	82.6	45.3	45.3	48.9	60.5	58.3
% vehicles left garages	87.4	49.2	49.4	53.5	63.1	61.1
Average travel time [s]	133.8	268.0	236.5	277.4	143.2	156.2
Average speed [mph]	13.5	6.4	7.1	6.1	11.6	11.3
Avg. delay time [s]	95.1	204.2	192.5	215.6	82.1	90.9
Avg. number of stops	3.2	5.1	4.9	5.8	3.2	3.6

Table 3. Core Area Performance through Simulation under Six Proposed Scenarios

	S1	S2	S3	S4	S5	S6
0 – 600 [s]						
# vehicles left area	3314	1206	1210	1238	1323	1419
% vehicles left garages inside the area	92.1	33.5	33.6	34.4	36.8	39.4
delay [s]	22.8	31.5	29.6	33.2	25.1	28.7
# stops	0.68	0.97	0.93	1.09	0.82	1.04
0 – 3600 [s]						
# vehicles left area	3796	1452	2201	1883	1551	1853
% vehicles left garages in core area	N/A	40.3	61.2	52.3	43.1	51.5
delay [s]	28.7	30.8	28.9	29.6	32.9	29.1
# stops	0.8	1.5	0.9	1.0	1.1	1.0

3 Evaluating Contra-Flow Plan Using Fuzzy Logic

3.1 Fuzzy Logic

Lofti A. Zadeh introduced the fuzzy set in 1965 at UC Berkeley [13]. Fuzzy logic is derived from fuzzy set theory dealing with reasoning that is approximate rather than precisely deduced from classical predicate logic. It can be thought of as the application side of fuzzy set theory dealing with well thought out real world expert values for a complex problem. Crisp set are first concerted into linguistic terms which is called fuzzification. The fuzzy set is mapped to another fuzzy set using rules or expert opinions. Then, the mapped fuzzy set is converted back to crisp set which is called defuzzification.

3.2 Using Fuzzy Logic to Evaluating Contra-Flow Plan

Contra-flow plan can be evaluated based on fuzzy logic. Before setting up fuzzy logic based evacuation evaluation plan, the inputs (MOEs from different scenarios) and outputs (the evaluation of a plan) of the fuzzy system should be firstly built up. Then through fuzzification, the crisp values will be converted into linguistic terms, and fuzzy rules will be tabulated. Finally the system output will be yield out.

3.2.1 Evaluating Contra-Flow Plan of Scenario 3 and 4

In table 3, the percentage of vehicle left the network in Scenario 3 is less than that in Scenario 4, while the other MOEs indicate contra-flow plan in Scenario 4 is better than Scenario 3. The following shows how fuzzy logic is used to synthesize these contradictory conclusions.

1)   Define the inputs and outputs
The three inputs and one output are defined as:

x_1 : Number of vehicles left the network

$x_1 \in \{\text{Not Acceptable (NA)}, \text{Acceptable (A)}, \text{Good(G)}\}$

x_2 : Average travel time (s) $x_2 \in \{\text{Short (S)}, \text{Acceptable (A)}, \text{Long (L)}\}$

x_3 : Average speed (mph) $x_3 \in \{\text{Low (L)}, \text{Acceptable (A)}\}$

y : Evaluation of contra-flow plan,

$y \in \{\text{Not Acceptable (NA)}, \text{Acceptable (A)}, \text{Good (G)}\}$

2)   Fuzzification

x_1 : Number of vehicles left the network (Fig. 3)

The number of vehicle 4602 is the result for Scenario 2 where no improvement is made, and is thus chosen as the worst case. While 60% and 90% of the capacity of 7 parking garages (10160), which are 6096 and 9050, are selected as the middle and good cases, respectively.

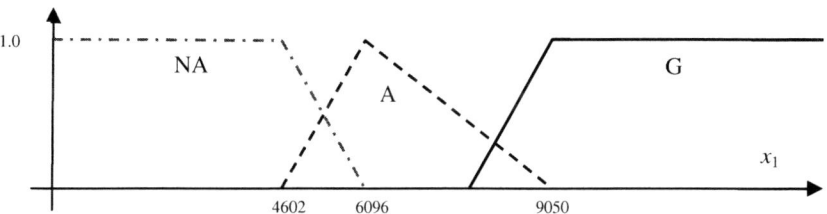

Fig. 3. Fuzzification of number of vehicles left the network x_1

x_2 : Average travel time (s) (Fig. 4)

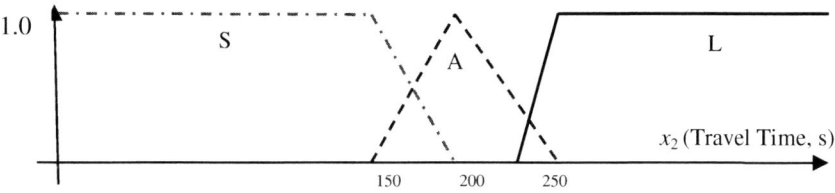

Fig. 4. Fuzzification of Average Travel Time x_2

x_3: Average speed (mph) (Fig. 5)

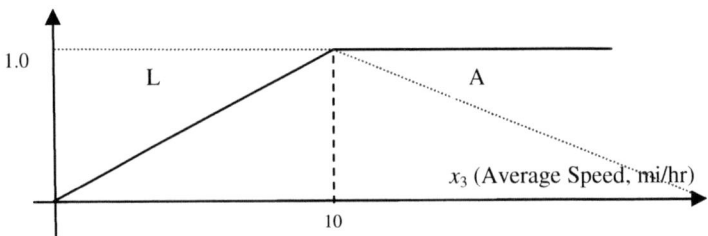

Fig. 5. Fuzzification of Average Speed x_3

3) Rules set up
The set-up rules are illustrated in Table 4.

Table 4. Rule for Evaluating Scenario 3, 4

X_1	X_2	X_3	Y	X_1	X_2	X_3	Y
NA	S	L	NA	A	A	A	A
NA	S	A	NA	A	L	L	A
NA	A	L	NA	A	L	A	A
NA	A	A	NA	G	S	L	G
NA	L	L	NA	G	S	A	G
NA	L	A	NA	G	A	L	G
A	S	L	A	G	A	A	G
A	S	A	G	G	L	L	A
A	A	L	A	G	L	A	A

4) Inference output from set-up rules
Based on Table 4 when applying the inputs for Scenario 3 and Scenario 4, the outputs are: $y_{scenario3}$ = NA, and $y_{scenario4}$ = A. So contra-flow operation in Mousund Street heading east (Scenario 4) is better than the plan on John Avenue heading south (Scenario 3).

3.2.2 Evaluating Contra-Flow Plan of Scenario 5 and 6
1) Define the inputs and outputs
   x_1 : Number of vehicles left the network
   $x_1 \in \{$Not Acceptable(NA), Acceptable(A), Good(G)$\}$
   x_2 : Number of vehicles left core area during 0-3600s
   $x_2 \in \{$Not Acceptable(NA), Acceptable(A), High(H)$\}$
   y : Evaluation of contra-flow plan
   $y \in \{$Not Acceptable(NA), Acceptable(A), Good(G)$\}$

2)   Fuzzification

    x_1 : Number of vehicles left the network

    x_2 : Number of vehicles left core area during 0-3600s

In Fig. 6, 1600 is the number of vehicles in Scenario 2 with no improvement, and 3250 is 90% of the total capacity of 4 parking garages inside the core area (3600). So 1600 and 3600 are selected as the two ends of the membership functions.

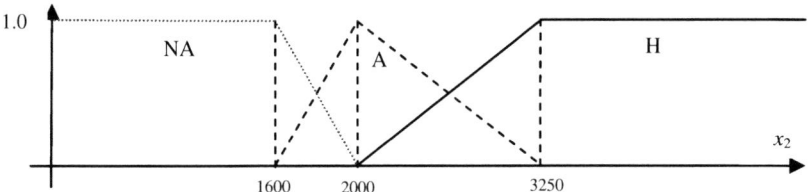

Fig. 6. Fuzzification of Number of Vehicle left the Core Area x_2

3)   Set up rules

Table 5. Rule for Evaluating Scenario 5, 6

X_1	X_2	Y	X_1	X_2	Y
NA	NA	NA	A	H	G
NA	A	NA	G	NA	NA
NA	H	NA	G	A	G
A	NA	NA	G	H	G
A	A	A			

5)   Inference output from set-up rules

Based on Table 5, $y_{\text{scenario 5}}$ = NA while $y_{\text{scenario6}}$ = A. So contra-flow operation in Holcombe Boulevard heading east is better than the plan heading west.

3.2.3   Evaluating Contra-Flow Plan of Scenario 3, 4, 5, 6

Then we try to assess four plans in a uniform way.

1)   Define the inputs and outputs

    x_1 : Number of vehicles left the network

    $x_1 \in$ {Not Acceptable(NA), Acceptable(A), Good(G)}

    x_2 : Average travel time (s) $x_2 \in$ {Short(S), Acceptable(A), Long(L)}

    x_3 : Number of vehicles left core area during 0-3600s

    $x_3 \in$ {Not Acceptable(NA), Acceptable(A), High(H)}

    y : Evaluation of contra-flow plan

      $y \in$ {Not Acceptable(NA), Acceptable(A), Good(G), Very Good(VG)}

2)   Fuzzification
   x_1 : Number of vehicles left the network (Fig. 3)
   x_2 : Average travel time (s)
   x_3 : Number of vehicles left core area during 0-3600s
3)   Set up rules (Table 6)

Table 6. Rule for Evaluating Scenario 3, 4, 5, 6

X_1	X_2	X_3	Y	X_1	X_2	X_3	Y	X_1	X_2	X_3	Y
NA	S	NA	NA	A	S	NA	NA	G	S	NA	NA
NA	S	A	NA	A	S	A	A	G	S	A	G
NA	S	H	NA	A	S	H	G	G	S	H	VG
NA	A	NA	NA	A	A	NA	NA	G	A	NA	NA
NA	A	A	NA	A	A	A	A	G	A	A	A
NA	A	H	NA	A	A	H	A	G	A	H	G
NA	L	NA	NA	A	L	NA	NA	G	L	NA	NA
NA	L	A	NA	A	L	A	A	G	L	A	A
NA	L	H	NA	A	L	H	A	G	L	H	A

4)   Inference output from rule table
From Table 6, $y_{scenario3}$ = NA, $y_{scenario4}$ = A, $y_{scenario5}$ = NA and $y_{scenario6}$ = A. Given all the rest of the MOEs are very close, as the travel time of Scenario 6 is much shorter than that of Scenario 4, so scenario 6 is considered as the best one among the four.

4   Conclusion

In this research, fuzzy logic is used to help evaluating the contradictory MOEs after simulation under different scenarios for contra-flow evacuation plan in Texas Medical Center area. While among the proposed four contra-flow plan, the one on Holcombe Blvd heading east is the best one, this may not be the firm optimal one since the combinations of all possible contra-flow countermeasures are listed as candidates. Therefore, a comprehensive study on this will be further conducted in order to provide an optimal evacuation plan for the study network. Further, more complicated rules as well as other operational countermeasures such as the optimal signal timing at intersections should also be considered in the future stage.

References

1. Seeba, J.: A Survey of Hurricane Katrina and Rita Contracts and Grants. US Department of Commerce, Final Audit Report No. DEN-17829-6-0001/July (2006)
2. Wolshon, B., Catarella-Michel, A., Lambert, L.: Louisiana Highway Evacuation Plan for Hurricane Katrina: Proactive Management of a Regional Evacuation. Journal of Transportation Engineering 1(132), 1–10 (2006)

3. Fantz, A., Meserve, J.: Witness survives by pretending to be dead. CNN News Report (April 17, 2007)
4. Alsnih, R., Stopher, P.R.: A Review of the Procedures Associated with Devising Emergency Evacuation Plans. In: The 83rd Transportation Research Board Annual Meeting, Washington, D. C. (January 2004)
5. U.S. Department of Homeland Security. National Response Plan: Nuclear/Radiological Incident Annex (December 2004)
6. Ardekani, S.A.: Management of Urban Road Networks Following Man-Made or Natural Disasters. In: Proceedings of the First International Conference on Safety and Security Engineering (SAFE 2005), pp. 541–546 (June 2005)
7. Department of Homeland Security: National Response Plan (December 2004)
8. Wolshon, B.: One-Way-Out: Contraflow Freeway Operation for Hurricane Evacuation. Natural Hazards Review 2(3), 105–112 (2001)
9. Tiglioglu, T.S.: Modeling Hurricane Evacuation Using Transportation Models, Fuzzy Set, and Possibility Theory, Texas Tech University
10. Poudel, M., Camino, F.M., de Coligny, M., Thiongly, J.A.: A Fuzzy Logic Approach for Aircraft Evacuation Modelling. In: 18th International Conference on Systems Engineering, ICSEng 2005, pp. 3–8, August 16-18 (2005)
11. Li, S., Xiao, G.: Fuzzy Synthetic Assessment on Safety Evacuation Measures in Library. Fire Science and Technology (January 2004)
12. VISSIM 4.10 Manual, PTV Inc. (2005)
13. Zadeh, L.A.: Fuzzy Sets. Information and Control 8, 338–353 (1965)

Arrow Exit Sign Placement on Highway Using Fuzzy Table Look-Up Scheme

Fengxiang Qiao, Xiaoyue Liu, and Lei Yu

Department of Transportation Studies, Texas Southern University, 3100 Cleburne Street,
Houston, TX 77004, USA,
Tel.: (713) 313-1915; Fax: (713) 313-1856
{qiao_fg,liux,yu_lx}@tsu.edu
http://transportation.tsu.edu

Abstract. In order to provide sufficient signing information to effectively guide drivers on highway, the advance guide sign should be wisely placed. This research employs a state-of-the-art/practice driving simulator to conduct experiment for evaluating the placement of arrow exit sign on highway taken various factors into consideration. The experimental design focuses on investigating traffic flow rate, highway geometric condition, and human behavior effect on sign placement. A fuzzy table look-up scheme is applied to build up the rule base of input-output pairs. A fuzzy logic system is constructed based on the outcome from the driving test and survey. Based on the analytical results, the optimal advance placement of arrow exit sign is recommended under different combinations of input variables.

Keywords: Fuzzy system, Table look-up Scheme, Sign placement, Arrow exit sign, and Driving simulator.

1 Background

Drivers, especially outlanders, rely heavily on guide signs to inform them of intersection or interchange approaches, to direct them correctly on freeways and streets, to identify upcoming exits towards destinations. The lack of signing information may engender drivers' confusion and even cause the safety problems.

The 2003 Manual on Uniform Traffic Control Devices [1], offers standards, guidance, and options for freeway guide signing, but no systematical methodology to determine sign placement. For example, for major and intermediate interchanges, it indicates that advance guide signs should be placed at 1 km (or 0.5 miles), and at 2 km (or 1 mile) in advance of the exit. However, there are no guidelines for installing guide signs within 0.5 miles from exits.

Li proposed a geometric relation based operational model for advance sign placement, [2]. However, that model is relatively static which did not involve dynamic variables such as traffic flow, and other non-technical factors (drivers' psychology, driving experience, etc.) to determine the sign placement.

The in-lab simulation, which adopts a state-of-art/practice driving simulator to conduct research on sign placement, can provide a dynamic driving environment to

E. Avineri et al. (Eds.): Applications of Soft Computing, ASC 52, pp. 260–269.

incorporate human behavior variables into model development. Studies have shown that driving simulator can accurately recreate driving conditions and, in turn, realistic driver behavior.

When considering the variables or parameters that affect sign placement determination, not all them are crisp sets. For example, it is hard to determine an absolute value of high and low speed, or congested and uncongested traffic flow. The whole system of identifying sign placement is difficult to be precisely described due to the human knowledge incorporation.

2   Objective

Currently, all the exit direction signs are installed at the exit point on highway. Due to complexity of real time traffic conditions, drivers may not have enough perception-reaction time to correctly make the lane-changing maneuvers towards the exit based on limited signing information. The objective of this research is to identify the feasibility of arrow exit sign placement under changing variables. Fuzzy inference system is employed to evaluate the necessity and feasibility of placing extra arrow exit signs on highway under various driving conditions.

3   Fuzzy Systems Using a Table Look-Up Scheme

Fuzzy logic theory, initiated by Lotfi A. Zadeh in 1965 with his seminal paper *"Fuzzy Sets"*, is capable of dealing with the systems where the precise descriptions are too complicated to be obtained and where the human knowledge is available to be combined into system [3]. Since the influences of the factors for determining the sign placement are so complicated and there do exist some human knowledge that need to be taken into consideration in sign placement, fuzzy system is a good candidate to construct the desired sign placement scheme.

Fuzzy table look-up scheme designs fuzzy systems from data sets (see, e.g. Wang [4]). It has been used in a variety of transportation applications including truck backer-upper control, time delay estimation at signalized intersections [5], freight transport assessment [6], etc.

Suppose that the following input-output pairs are provided:

$$\left(x_1^p, x_2^p, \ldots, x_n^p; y^p \right), p = 1, 2, \ldots, N \tag{1}$$

where, $\left(x_1^p, x_2^p, \ldots, x_n^p \right) \in U = [\alpha_1, \beta_1] \times [\alpha_2, \beta_2] \times \ldots \times [\alpha_n, \beta_n] \subset R^n$ are all the considered input variables to the sign placement model, while $y_0^p \in V = [v_y, \omega_y] \subset R$ is the output place to have the arrow exit sign. The input variables $\left(x_1^p, x_2^p, \ldots, x_n^p \right)$ may be the traffic flow rate, speed when driver sees the sign, speed when driver is exiting, number of lanes, etc. The proposed fuzzy system should be based on the rule base generated from these N input-output pairs. The following five-step scheme can be used to design the fuzzy system [2].

3.1   Define the Fuzzy Sets to Cover the Input and Output Spaces

Specifically, for each $[\alpha_i, \beta_i]$ $i = 1, 2, ..., n$, define N_i fuzzy sets A_i^j $(j = 1, 2, ..., N_i)$, which are required to be complete in $[\alpha_i, \beta_i]$, that is, for any $x_i \in [\alpha_i, \beta_i]$, there exists A_i^j such that its membership values $\mu_{A_i^j}(x) \neq 0$. For example, the pseudo-trapezoid membership functions are possible candidates.

3.2   Generate One Rule from One Input-Output Pair

First, for each input-output pair $(x_1^p, x_2^p, ... , x_n^p; y^p)$, determine the membership values of x_i^p $(i = 1, 2, ..., N_i)$ in fuzzy sets $A_i^j (j = 1, 2, ..., N_i)$ and the membership values of y^p in fuzzy sets B^l $(l = 1, 2, ..., N_y)$. That is, compute $\mu_{A_i^j}(x_i^p)$ for $j = 1, 2, ..., N_i$, $i = 1, 2, ..., n$, $\mu_{B^l}(y^p)$ and for $l = 1, 2, ..., N_y$. Then for each input variable x_i $(i = 1, 2, ..., n)$, determine the fuzzy set in which x_i^p has the largest membership value, that determine A_i^{j*} such that $\mu_{A_i^{j*}}(x_i^p) \geq \mu_{A_i^j}(x_i^p)$, $j = 1, 2, ..., N_i$.   Similarly,   determine B^{l*} such   that $\mu_{B^{l*}}(y^p) \geq \mu_{B^l}(y^p)$, $l = 1, 2, ..., N_y$. Then we can obtain a fuzzy IF-THEN rule.

$$Ru^{(l)}:\ IF\ x_1\ is\ A_1^l\ and\ ...\ and\ x_n\ is\ A_n^l,\ THEN\ y\ is\ B^l \tag{2}$$

3.3   Assign a Degree to Each Rule Generated in 3.2

There are possibly conflicting rules existed, a degree is assigned to each rule so that only one rule is kept that has the maximum degree. The degree of the rules is determined by the reliability of associated input-output pairs. Therefore providing the input-output pair $(\mathbf{x}^p; y^p)$ has a reliable degree $\mu^p \in [0, 1]$, the degree of the rule generated by $(\mathbf{x}^p; y^p)$ is calculated by:

$$D(rule) = \prod_{i=1}^{n} \mu_{A_i^{j*}}(x_i^p)\ \mu_{B^{l*}}(y^p)\ \mu^p \tag{3}$$

3.4   Create the Fuzzy Rule Base

Possible linguistic rules from human experts (due to conscious knowledge), together with the rules from the input-output pairs will generate the desired rule base, which forms a look-up table. The entire process can be viewed as a table look-up scheme.

3.5   Construct the Fuzzy System Based on the Fuzzy Rule Base

Based on the rule base in 3.4, a fuzzy system can be eventually constructed with the product inference engine, singleton fuzzifier, and centre average defuzzifier:

$$f(x) = \frac{\sum_{l=1}^{M} \bar{y}^l \left(\prod_{i=1}^{n} \mu_{A_i^l}(x_i) \right)}{\sum_{l=1}^{M} \left(\prod_{i=1}^{n} \mu_{A_i^l}(x_i) \right)} \tag{4}$$

where, $\overline{y}^l$ is the centre value of the fuzzy set B^l in the output region, that is, the THEN part, for the l-th rule, $\mu_{A_i^l}(x_i)$ is the membership function of the lth rule for the ith component of the input vector. M is the total number of fuzzy rules.

Therefore, once a rule base is produced, the fuzzy system will yield out the desired placement of arrow exit sign corresponding to the input $x = (x_1, x_2, ... x_n)$, which may be the traffic flow rate, speed when driver sees the sign, speed when driver exits, number of lanes, etc.

4   Fuzzy Logic Based Arrow Sign Placement Procedure

The proposed methodology to determine the placement of extra arrow exit sign along highways is to simulate the real driving environment using driving simulator (See Fig. 1) and let the subjects test the scenarios of sign placement alternatives under variable traffic and geometric situations. After conducted the experimental test, a survey is required to obtain information of subjects' feeling and opinion concerning different placement candidates [7].

Fig. 1. TSU owned driving simulator DriveSafety DS-600c that was used for creating Fuzzy rule base from input-output data pairs

Although subjects can demonstrate and tell their preference through the driving simulation and survey, the subconscious knowledge, or say, underlying rules to determine the sign placement is hard to express in linguistic terms. Thus, when the driving behavior is demonstrating, the whole process is viewed as a black box with input-output data pairs generated. So forth, part of the rule base can be created.

4.1   Procedure of Fuzzy Based Arrow Exit Sign Placement

Step 1. Micro-Simulation Process. In order to achieve the utmost sense of reality, the scenarios in driving simulator should generate initial vehicles around the subject vehicle with initial speed, and with determined traffic flow rate. The vehicle movements should also be considered to control the vehicles' interaction.

Step 2. Scenario Design. To replicate the actual driving environment, the scenario design should build the freeway structure that the research intends to simulate, the background scene, the sign types and other advance guide signs besides the arrow exit sign.

Step 3. Subject Testing. Subjects are recruited to conduct test on driving simulator. They will be told to arrive a destination along highway based on the guide signing information. Drivers have control of the acceleration pedal, brake pedal, and the steering wheel, exactly like they do in a real automobile. Subjects will also be told to press a programmable button when they see the arrow exit sign clearly. Data about driver position, velocity, brake pressure and the time they press the button can be collected in real time.

Step 4. Survey and Questionnaire. After subjects conduct the test, they will finish a survey concerning their subjective evaluation of arrow exit sign placement under various traffic and highway conditions. The survey result will further be analyzed and combined with testing outcome.

Step 5. Fuzzy Rule Base Establishment. After collect the data from objective testing and subjective questionnaire, the information will be further delved. For each combination of the input variables, there will be an output. Input and output crisp data will be allocated to one or two fuzzy sets with associated membership values. By relating fuzzy sets related to input-output data pairs, rules from the experiment and survey can be obtained one by one. The synthesized rule base is later merged from the combination of experiment and survey, with the experimental outcome being the principle part and the survey result as a supplement.

Step 6. Construct Fuzzy System and Determine Sign Placement. With the created rule base and the defined fuzzy sets covering input and output spaces, the fuzzy system is constructed. With this established system, sign placement under various combinations of inputs sampled within the range of the input space, will be inferred.

5 Case Study

As an illustration, the case study demonstrates a procedure using both driving simulator experiment and posterior survey to determine the placement of exit arrow sign along highway.

5.1 Micro-simulation Process

The distribution of the headways between the vehicles generated follows negative exponential distribution [8]. Specifically speaking, the vehicles will be generated at an interval h, calculated by the following equation:

$$h = (H_{avg} - h_{min})[-\ln(1-R)] + H_{avg} - h_{min} \tag{5}$$

where, h is the time interval; H_{avg} is the average headway; h_{min} is the minimum headway; and R is a random number in the range of 0.0 to 1.0.

The vehicle type on highway is also an important parameter. The percentage of heavy vehicles is counted via video from U.S. Highway 101 (Hollywood Freeway) in Los Angeles, California collected between 8:05am and 8:20am on June 15, 2005 [8].

The vehicle movements are controlled by a dynamic model, which consists of two parts. One is the vehicles' accelerations determination for each frame; the other is the lane change model.

5.2 Experiment Description

Three scenarios with different number of freeway lanes are designed for the experiment (See Fig. 2).

In Fig. 2, the freeway types for all the scenarios are the same, that is, one mainline with one exit. The default scenarios all have guide signing information of Almeda Rd. exit with advance guide sign in 1 mile, 1/2 mile, and at the exit point.

For all the scenarios, the speed limit is 65mph on mainline, and 45 mph on the exit ramp. Two kinds of traffic flow rate will be tested during simulation run. One is of Level of Service[1] B (LOS B), which is in represent of uncongested situation, the other one is of Level of Service D (LOS D), which is considered as congested situation.

For each of the scenario, there are three alternatives of arrow exit sign placement. First is the one only installed at the exit point, second is an extra arrow exit sign installed near the exit (at 1/4 mile from the exit), third is an extra arrow exit sign installed further from the exit (at 3/8 mile from the exit). Thus, each scenario will have three sign placement situations.

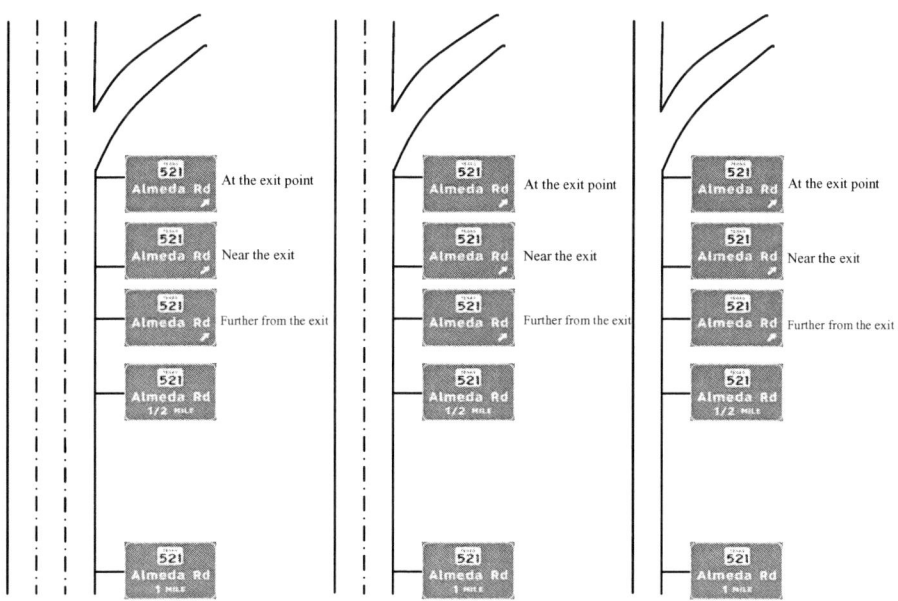

Fig. 2. Scenario Design

---

[1] Level of Service (LOS) criteria is defined as six levels of traffic flow conditions from level A to F with A as the free flow status, and F the jam status. LOS B and LOS D correspond to a vehicle to capacity ratio of 0.49 and 0.9, respectively [9].

It was hypothesized that the visibility of each sign would vary as a function of the lane in which the driver was located and the cognitive demands on the driver. In order to determine the effect of these factors on driver performance, the following items were indicated:

1. The driver's starting lane - The initial position of driver is the leftmost lane, while the tested signs are all indicating the exit on the right.
2. Cognitive demands - The subjects are told to press the programming button when they think they clearly see the guide signs.

Forty-eight test subjects participated in this experiment (ages 23-45, 36 male, 12 female), and were divided into 3 groups. Each group tested one scenario shown on Fig. 2. Participants were instructed before experiment to exit at Almeda Rd. All participants were not familiar with the exit sign sequence or geometry. The number of missed exits was recorded, and the position when subjects pressed the button was recorded. Subjects were required to take a survey form after they conducted the test.

5.3 Test Result Analysis

The result of test run was recorded by the driving simulator system. For each combination of traffic flow rate, highway geometry, and driving behavior, every alternative

Table 1. Description of input variables[2]

Variables	Definition	Levels	Terms			Centers		
U	Average speed when seeing the sign	3	L	M	H	35	50	65
V	Average speed when exiting	3	L	M	H	20	30	45
F	Heavy flow or light flow	2	B	D		B	D	
N	Number of lanes	3	1	2	3	1	2	3
P	Sign position	3	A	N	F	0.000	0.250	0.375

Table 2. Rule developed by test

			Rules		
No.	If U is	And if V is	And if F is	And if N is	Then P is
1	L	L	B	1	A
2	L	L	B	2	N
3	L	L	B	3	N
4	L	L	D	1	N
5	L	L	D	2	N
6	L	L	D	3	F
	...	...	...		
	...	...	...		
	...				
50	H	H	B	2	F
51	H	H	B	3	F
52	H	H	D	1	N
53	H	H	D	2	N/A
54	H	H	D	3	N/A

[2] Terms: L=low, M=middle, H=high; A=at the exit point, N=near the exit, F= further from the exit. Unit of centre value: U, V – mph; P– mile.

will be tested and analyzed to find the optimal one for each condition. The criteria include error percentage; lane changing maneuver after driver saw the sign; speed; and the sign placement effect on driver's visual acuity.

Incorporating all these information collected from the simulation, the optimal alternatives under each combination of variables are listed in Table 1 and 2.

The result of survey was also analyzed in the format as Table 3.

Table 3. Rule developed by survey

No.	If U is	And if V is	And if F is	And if N is	Then P is
			Rules		
1	L	L	B	1	A
2	L	L	B	2	N
3	L	L	B	3	A
4	L	L	D	1	F
5	L	L	D	2	N
6	L	L	D	3	N
	...		...		
	...		...		
	...		...		
50	H	H	B	2	F
51	H	H	B	3	N
52	H	H	D	1	A
53	H	H	D	2	N
54	H	H	D	3	N

A synthesized rule is further developed based on the rules from test and survey. The synthesized one considers the test result as principle part, with survey result as a supplement when the combination of items is vacant in test (See Table 4). Also, the membership functions of input variables are demonstrated at the top of the table.

Table 4. Synthesized rule of test and survey

No.	If U is	And if V is	And if F is	And if N is	Then P is	Center
1	L	L	B	1	A	0.000
2	L	L	B	2	N	0.250
3	L	L	B	3	N	0.250
4	L	L	D	1	N	0.250
5	L	L	D	2	N	0.250
6	L	L	D	3	F	0.375
		...	...			
		...	...			
		...	...			
50	H	H	B	2	F	0.375
51	H	H	B	3	F	0.375
52	H	H	D	1	N	0.250
53	H	H	D	2	N	0.250
54	H	H	D	3	N	0.250

Except for variable of number of lanes which used singleton membership function, all the other variables adopted pseudo-trapezoid membership functions.

5.4 Fuzzy System Establishment

The rule base created in Table 4 covers all the centre value situation of variables. However, for some value point (for example, 40mph speed when subjects saw the sign), there is no corresponding optimal sign placement result. Thus, we need to establish a fuzzy system to provide a complete database.

We use the fuzzy system with product inference engine, singleton fuzzifier, and center average defuzzifier; that is, the designed fuzzy system in the form of Equation (4) with the rules in Table 4.

As an illustration, Table 5 shows part of the output with a speed increment of 5 mph using the fuzzy system. It demonstrates the optimal sign placement for highway with 3 lanes.

Table 5. Fuzzy Inference Engine of Optimal Sign placement for 3-Lane Highway

D3		Sign Position for Heavy Traffic with 3 Lanes						B3		Sign Position for Light Traffic with 3 Lanes					
		20	25	30	35	40	45			20	25	30	35	40	45
L	35	0.375	0.375	0.250	X	X	X	L	35	0.250	0.250	0.000	X	X	X
	40	0.375	0.375	0.250	0.250	X	X		40	0.250	0.250	0.000	0.000	X	X
	45	0.375	0.375	0.375	0.375	0.375	X		45	0.250	0.250	0.250	0.250	0.250	X
M	50	0.375	0.375	0.375	0.375	0.375	0.375	M	50	0.250	0.250	0.250	0.250	0.250	0.250
	55	0.375	0.375	0.375	0.375	0.375	0.375		55	0.250	0.250	0.250	0.250	0.250	0.250
	60	0.375	0.375	0.000	0.000	0.250	0.250		60	0.375	0.375	0.250	0.250	0.375	0.375
H	65	0.375	0.375	0.000	0.000	0.250	0.250	H	65	0.375	0.375	0.250	0.250	0.375	0.375
		L		M		H				L		M		H	

6 Conclusion

In this paper, an experiment to determine the placement of arrow exit sign is conducted and a fuzzy logic based arrow sign placement approach is proposed.

The experiment that utilized driving simulator as a tool to evaluate driver's behavior on sign placement takes human knowledge into consideration, which successfully replicates the real driving situation and cognitive driving behavior. The fuzzy logic based sign placement approach converts the linguistic human knowledge, combining with the vehicle trajectory information, into a mathematical system to determine the optimal placement of arrow exit sign along highway.

The case study of the paper demonstrated the procedure to determine the optimal placement of arrow exit sign of a highway with different number of lanes (1, 2 and 3) under various traffic conditions. A fuzzy system is established using a table look-up scheme which combines technical and non-technical factors.

This approach shows that fuzzy systems can be used as a tool to find the optimal result in determining the sign placement. It can also be implemented in sign design evaluation in a wider application area.

Acknowledgement

This research is sponsored by Texas Department of Transportation research project 0-5800. The authors would like to thank project director Ismael Soto, and all others who contributed to this research.

References

1. Manual on Uniform Traffic Control Devices for Streets and Highways, adopted by the Federal Highway Administration (2003)
2. Li, J., Lan, C., Chimba, D.: A Supplement to Advance Guide Sign Placement Guidelines in MUTCD. In: Transportation Research Board 85th Annual Meeting compendium of papers, CD-ROM. Transportation Research Board of the National Academies, Washington, D.C (2006)
3. Zadeh, L.A.: Fuzzy Sets. Information and Control 8, 338–353 (1965)
4. Wang, L.X., Mendel, J.M.: Generating Fuzzy Rules by Learning from Examples. IEEE Transactions on Systems, Man, and Cybernetics 22(6) (1992)
5. Qiao, F., Yi, P., Yang, H., Devaradonda, S.: Fuzzy Logic Based Intersection Delay Estimation. Journal of Mathematical and Computer Modelling 36(11) (2002)
6. Qiao, F.G., Lei, X., Yu, L.: Neuro-Fuzzy Logic Based Model for Assessment of Freight Transportation Implications on Communities. In: Transportation Research Board 80th Annual Meeting compendium of papers, CD-ROM. Transportation Research Board of the National Academies, Washington, D.C. (2001)
7. HyperDrive & Vection User's Guide Version 1.9.35, DriveSafety Inc. (2005)
8. Federal Highway Administration (2007), http://www.ngsim.fhwa.dot.gov/
9. Transportation Research Board. Highway Capacity Manual. National Research Council, Washington, D.C. (2000)

Index

Author Index